AF882738

COMPLETE YEAR 7

NAPLAN

Copyright © 2023 MAYANK KUMAR

The rights of Mayank Kumar to be identified as the author of the work has been asserted by him in accordance with the Copyright Act of Australia, 1968.

All rights reserved.

No part of this book may be reproduced or transmitted in any form or by any means, electronic or mechanical, including photocopying, recording or by any information storage and retrieval system, without prior permission in writing from the publisher.

For any enquiries or permissions, the publisher can be contacted at mk3265050@gmail.com

First published in Australia by Mayank Kumar

First Print: 2023

DISCLAIMER

This book, titled "[Book Title]: A Guide to Mathematics Solutions," is an independent educational resource created solely for the purpose of aiding students in understanding mathematical concepts and problem-solving techniques. The author of this book is in no way affiliated or connected with the National Assessment Program – Literacy and Numeracy (Naplan) or any educational institution.

The solutions provided in this book are intended to serve as instructional tools to assist students in comprehending and mastering mathematical problems similar to those encountered in Naplan assessments. They are not endorsed, sponsored, or approved by Naplan or any associated educational authorities.

Readers are encouraged to use this book as a supplementary resource to enhance their mathematical skills and gain a deeper understanding of problem-solving techniques. However, it is important to note that this book does not guarantee any specific outcomes on Naplan assessments or any other standardized tests.

The author and publisher of this book have made every effort to ensure the accuracy and clarity of the solutions provided. Nevertheless, readers are advised to consult official educational resources and guidelines for the most up-to-date information and assessment requirements.

By using this book, readers acknowledge that the author and publisher are not responsible for any consequences, outcomes, or interpretations related to Naplan assessments or any other examinations. This book is intended solely as a support tool for educational purposes.

Please note that the author's primary goal is to foster a deeper understanding of mathematics among students and to provide valuable assistance in their academic journey.

WORKBOOK AND TESTS
NUMERACY STEP BY STEP SOLUTIONS

WRITER – MAYANK KUMAR

BEGINNER NUMBER

NUMBER LINE

SOLUTION 1: Annual fees for Jody to own a netball club = $ 14

Additionally, she buys two caps for = $ 12

But Jody only pays (in total) = $ 10

Now, the question is asking how much (in dollars) she still owes to the club.

(HINT: Add she is required to pay to the club (in total) and subtract the amount

(In dollars) that she has already paid to the club).

Therefore, 14 + 12 – 10 = 26 – 10 = 16 dollars

Thus, 16 dollars is the amount that she STILL owes to the Netball Club.

ALTERNATIVE SOLUTION: USING NUMBER LINE

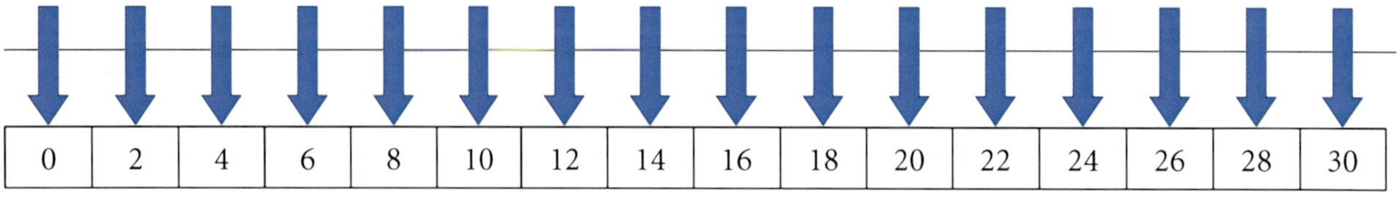

(in dollars)

Annual fees of the Netball Club = Equals 14 dollars

Cost of 2 caps = Equals 12 dollars

So, Adding these two values = 14 + 12 = 26 dollars

However, Jody pays 10 dollars to the club, thus we subtract 10 dollars from the opposite side.

Jody paid it already = 10 dollars

Therefore, the amount that Jody still owes to the club is 26 – 10 = 16 dollars

SOLUTION 2: Kris punted the football by = 35 metres

Dean forced the ball back to Kris by = 40 metres

Remember, the actual initial position of Kris can be literally anywhere on the number line, that is the position he punted the ball towards Dean.

To calculate : The distance Kris would run to kick the ball

Reason, we need to calculate the distance in metres, so it does not really matter where the POSITIONS are of Kris and Dean.

So, let's say Kris, initially, is at number 5 on the number line.

Now, using number line, let's find out how much distance (in metres) Kris need to run back to kick the ball.

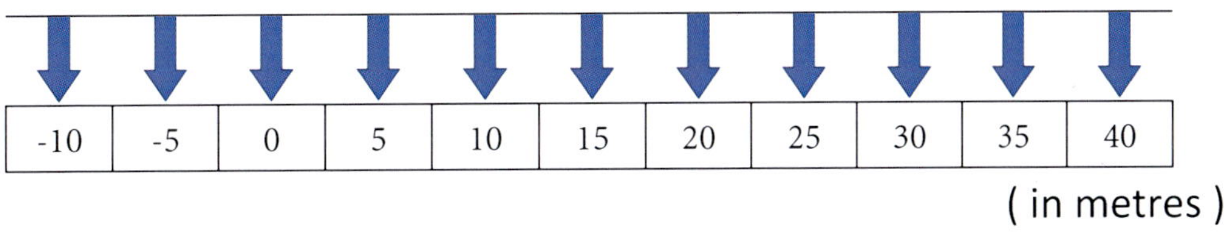

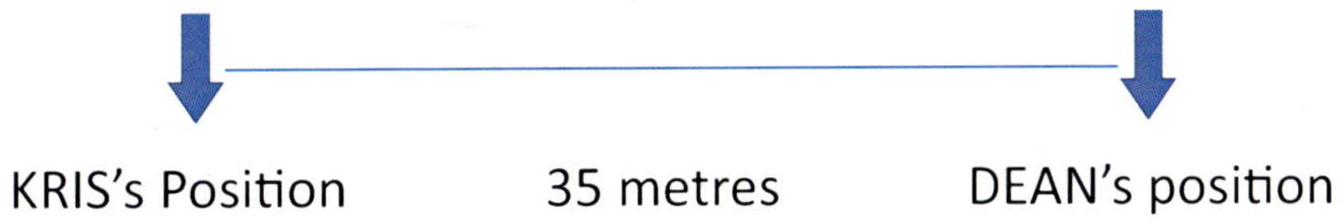

When Dean Kicked it back, the scenario changes to

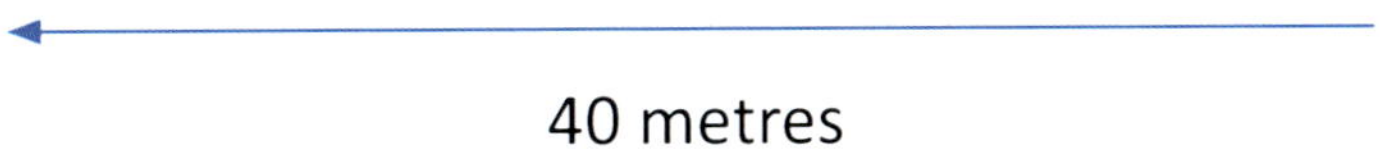

Now, Kris will need to run back from his actual position to kick the ball would be

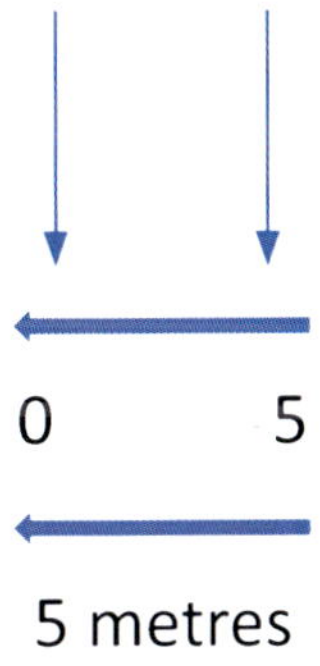

So, Kris needs to run 5 metres.

LARGE NUMBERS

SOLUTION 1: To find a number that is closest to three- quarters of a million, we find out how much it is numerically.

1 million: It is a number equal to one thousand thousands (1,000,000).

To calculate its three quarter (3/4), we multiply it by ¾ * 1000000 = 750000.

Therefore, we need to find out which option is closest to 750000.

In the given options, the number is closest to 750000.

Now, have a glance at option number 2, it is closest to 750000 because it has the least difference of 2110 among all 4 options provided.

So, the option (B) 752110 is correct.

Write the following numbers in figures.

FUN FACT: REMEMBER THE PLACE VALUE SYSTEM TABLE:

And so on.......	Hundred-Millions	Ten-Millions	One-Millions	Hundred-Thousands	Ten-Thousands	One-Thousands	Hundreds	Tens	Ones

SOLUTION 2: USING THE ABOVE TABLE,

We have got '950' thousands and having space for hundreds, tens, and ones (3 digits after 950)

Similarly, question says we have Hundreds = 9 means leaving space for 2 digits after 9 i.e., tens and ones

Additionally, as per the question states, we add 4.

So, we can write it as 950,904.

SOLUTION 3: USING THE ABOVE TABLE,

Five million = 5_ _ _ _ _ _

Thousands = 109 _ _ _

Hundreds = 0

Tens and ones = 77

So, the number will be 5,109,077

SOLUTION 4: USING THE ABOVE TABLE, we conclude that

Thousands = 80 so it becomes 80 _ _ _

Hundreds = 3 so it becomes 80,3_ _

Tens = 3

Ones = 0

Thus, the final answer is, using all these equations will be 80, 330.

PLACE VALUE

SOLUTION 1: We must write the place value of 5 in 54,299

Using the PLACE VALUE SYSTEM TABLE, WE CALCULATE FROM THE END SIDE ← OF THE GIVEN NUMBER, means the last digit, 9 is at one's place

Again, 9 is at tens place as well

Furthermore, 2 is at hundreds place.

Last but not the least, talking about 4, it is at thousands place.

Finally, we reach the very last digit, that is 5, will be at Tens Thousands place.

SOLUTION 2: Similarly, finding out the place value of 5 in 793548.

Repeating the steps that have been used in last Question, 5 will be at Hundreds place (THIRD PLACE VALUE) if we count from RIGHT TO LEFT ←.

SOLUTION 3: Finding the PLACE VALUE of 5 in the given number 75203,
←

3 has got a place value of "ones"

0 has got a place value of "tens"

2 has got a place value of "hundreds"

Similarly, 5 will be having a place value of "thousands"

Thus, the answer to this question is Thousands.

SOLUTION 4: We must find the place value of 5 in the given number '8952'.

If we begin to count from RIGHT TO LEFT, 2 will hold a 'ones' place

Going further, the number '5' will hold a 'tens' place According to the PLACE VALUE SYSTEM.

MULTIPLICATION AND DIVISION

Write answers to the following.

1. 420
   X 9

   3780

STEPS TO DO:

Step 1: Multiply 9 and 0, that gives us 0

Step 2: Multiply 9 and 2, that gives us 18, We write down 8 and we will use 1 as a CARRY.

Step 3: Now, multiplying 9 and 4, that gives the numerical value of 36.

Final Step: Add the carry to 36, finally it becomes 36+1 = 37.

So, multiplying 420 by 9 gives us 3780.

2. 8 3
   X 20

   00
   166 x

   1660

STEPS TO DO:

NOTE DOWN: This type of example is a multiplication of two-digit numbers TO two-digit numbers.

Step 1: Multiply 0 and 3 which gives us 0

Step 2: Multiply 0 and 8 which gives us 0

Step 3: Put a cross mark under the first row, it is a sign of multiplication

Step 4: Repeat the process of multiplying, now we multiply 2 and 3, it gives us 6

Step 4: Finally, last stage of multiplication, we will multiply 2 and 8, as a result it gives us 16.

NOTE DOWN: The CARRY IS 1, but we will ignore that.

REASON: There are no quantities left for multiplication, so the CARRY 1 will automatically add up to make it 16.

Thus, the result is 1660.

3. 203
   X 5

1015

STEPS TO DO:

Step 1: Multiply 5 and 3, it gives us 15. But we cannot write both the numbers, therefore we will write the last digit i.e., 5 and '1' will be the CARRY.

Step 2: We add the CARRY to this step, here we multiply 5 and 0, the result is zero but adding CARRY 1 and zero will give us 1. So, write down 1.

Final step: Multiply 5 and 2, which gives us 10.

Therefore, the solution to this problem is 10.

4. 180 ÷ 3

   OR $\frac{180}{3} = 60$ 180 is an absolute divisor of 3 and leaves remainder 0

5. 2808 ÷ 12

   OR $\frac{2808}{12} = 234$ 2808 is an absolute divisor of 12 and leaves remainder 0

SOLUTION 6: To find a number that is halfway to any given two numbers will be the AVERAGE of those two numbers.

METHOD TO CALCULATE AVERAGE OF TWO NUMBERS: ADD BOTH NUMBERS AND DIVIDE BY 2!

Here, the two numbers are 18 and 32.

SPEP 1: ADD 18 and 32, the result is 50. Divide 50 by 2, it gives us 25.

Therefore, the result is 25.

SOLUTION 7: Using the above steps to calculate HALFWAY or AVERAGE of two numbers,

Add -10 and 30, By reversing it we can say add 30 and -10 OR we can re-write it as:

$$30 + (-10) = 30 - 10 = 20$$

Final Step: Now, divide 20 by 2 = 10

Therefore, the answer is 10.

NEGATIVE NUMBERS

SOLUTION 1: Given that : Dead sea in metres that is below sea level = 420 m approximately

Lake Titicaca in meters that is above sea level = 3810 m

To calculate : Height difference between Lake Titicaca and Dead sea (in meters)

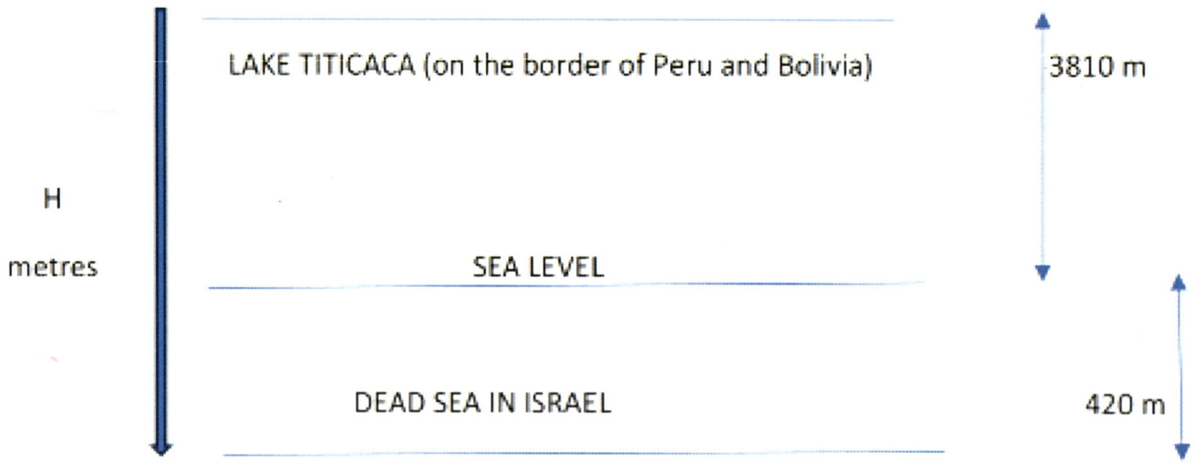

From the above diagram, we clearly see that Lake Titicaca is above 3810 metres above the sea level.

Similarly, the Dead Sea is below 420 metres below the sea level.

We need to find the height difference in metres between Lake Titicaca and the dead sea, which is the total difference.

Let's say the total difference = H

Therefore, H = 3810 metres + 420 metres = 4230 metres

So, the height difference in metres is 4230 metres.

SOLUTION 2: Given that : Tess's uncle has got in his savings account = $ 2396

Money must be transferred to the travel agent = $ 3427 .... (given)

The cost of Rucksack is = $ 250 .... (given)

So, the total amount of money that must be taken out from savings amount is = Cost of travel agent + cost of rucksack = $ 3427 + $ 250 = $ 3677

But Tess's uncle has only $ 2396 in his savings account.

So, the smallest amount of money he will need is $ 3677 - $ 2396 = $ 1281

Therefore, Tess's uncle needs to put 1281 dollars into his savings account to pay for the travel agent and the rucksack.

BEGINNER NUMBER
FRACTIONS AND DECIMALS

Complete the following tables:

FRACTION	DECIMAL
1/10	0.1
1/5	
	0.25
½	

FRACTION	DECIMAL
	0.6
3/4	
4/5	
	0.9

Let's go through each to find their solution,

* 1/5 = 0.2	(Converting from Fraction to Decimal)
* 0.25 = 25 / 100 = 1/4 (in shortest form)	(Converting from Decimal to Fraction)
* 1/2 = 0.5	(Converting from Fraction to Decimal)
* 0.6 = 6/10 = 3 /5 (in shortest form)	(Converting from Decimal to Fraction)
* 3/4 = 0.75	(Converting from Fraction to Decimal)
* 4/5 = 0.8	(Converting from Fraction to Decimal)
* 0.9 = 9/10 (in shortest form)	(Converting from Decimal to Fraction)

Now, putting the solutions back into the table.

FRACTION	DECIMAL
1/10	0.1
1/5	0.2
1/4	0.25
½	0.5

FRACTION	DECIMAL
6/10	0.6
3/4	0.75
4/5	0.8
9/10	0.9

SOLUTION: Number of sprinters used in 200-m event = 4

The respective times of the race are as : 29.21 , 29.86 , 28.92 , 29.74

We need to find the time taken by Millie to finish the 200-m race.

It is being advised in the question that Millie came 4th in the race, so the time taken by Millie is the most time taken out of given times.

As we clearly see that the 29.86 is the most time taken, it means this is the time taken by Millie.

Therefore, the Millie took the highest time to finish the race, that is 29.86 seconds.

PROBLEM SOLVING

POINT TO REMEMBER :

[Fewer = Lesser]

SOLUTION: In this question, the given statements are related to each other.

Now, Midge broke guitar strings = 27

As per the question, Farah holds 5 times as many dollars as Midge has had broken strings,

Therefore, Farah holds = (5 x 27) dollars = 135

Now, Farah has 135 dollars in the bank account.

Davo has caught 12 fishes fewer than one-third of 135 (number of dollars that Farah has) One third of 135 is 135/3 = 45

= 45 - 12 = 33

Thus, Davo caught 33 fishes.

Now as Tilly is related to Davo as Tilly's longest one day bike ride is 97 more than Davo's fish haul.

Using this statement to find out, how many kilometres Tilly's one day bike ride = 97 + 33= 130 kilometres

Now, as we know Tilly's longest bike ride is 130 kilometres.

Moving ahead, Ally's latest count of online friends = 15 lesser than ½ of Tilly's longest bike ride

It means, 1/2 of 130 = 65 kilometres

Now, the final step says to reduce 15 from the Ally's longest ride., that is what Ally has online friends.

Hence , Ally has = 65 - 15 = 50 online friends.

So, 50 is the correct answer.

BEGINNER MEASUREMENT AND SPACE

LENGTH AND AREA

To calculate : Best length unit for below measurements

1. (HINT: The pencil is always in centimetres)
2. (HINT: A nail (thumb) is usually too small and in millimeters)
3. (HINT: 15 minutes' drive on cycle can cover the distance in kilometres)
4. (HINT: An oval is a typically a playground that is in metres)

Using the above hints, we conclude that the solutions of above questions are

1. Cm (centimetres)
2. Mm (millimeters)
3. Km (kilometres)
4. M (Metres)

SOLUTION : Given that : Jerry paints in a garage = 2 side walls + rear wall
Also , Jerry can buy the paint in tins that are either = 2-L & 500-mL (or ½ -L)

FUN FACT: A garage is usually in the form of A CUBOID. A CUBOID is solid figure that has six rectangular faces at right angles to each other.

Clearly seen from the question, we have been given the length, breadth, and height of the garage.

Let us assume, Length is L metres.

Breadth is B metres.

Similarly, Height is H metres

Therefore, Length, L = 6 metres

Breadth, B = 3 meters

Height, H = 2 metres

HOT TIP :
Length means how lengthy the garage is.
Similarly, Breadth means how wide the garage is.
Similarly, Height means how high is the Garage .

As per the question, Jerry only paints the inside of the garage that includes,

[Two side walls and the Rear wall]

Now, we will calculate the areas of these walls.

Area of 2 side walls = L x B + L x B = (6 x 2) + (6 x 2) = 24 m^2

FUN FACT:

The Garage has 6 rectangular faces, meaning the 4 walls, 1 roof and 1 floor has rectangle shape.

Therefore, the area of each face is (length x breadth)

POINT TO REMEMBER:

A rear wall is the wall that is opposite to the front wall of the garage.

ALWAYS REMEMBER: The side walls, the front wall and the rear wall have two dimensions (length and breadth) and as these are in Rectangular shape, therefore , these will have the AREA of the rectangle, which is (LENGTH x BREADTH).

In this case, the breadth of side wall will be itself the HEIGHT OF THE CUBOID (as the dimension changes)

Now, calculating the area of rear wall = (Length of rear wall) x (Breadth of rear wall)

So, AREA = 3 m x 2 m= 6 m^2

Total area that Jerry paints = 24 m^2 + 6 m^2 = 30 m^2

NOTE DOWN:

Area is always calculated in m^2 as it will have only two dimensions

$\therefore$ A= $(\text{dimensions})^2$

He can buy the paints in 2 liters or in 500 ml tins.

1 L of paint can cover = 14 m^2 (as per the question)

Therefore, 30 m^2 of the area can be painted with the help of at least a 2-L tin, because 2-L tin will contain sufficient paint to cover 28 m^2.

Out of 30 m^2, 28 m^2 is already covered and still 2 m^2 area to paint is left.

If 1 litre will cover 14 m^2,

[if 1 L covers 14 m^2, so 2 L covers = 28 m^2]

Therefore, ½ litre or 500 ml will cover 7 m^2 (□ 14 m^2/2)

To cover the remaining 2 m^2, we can buy another 500-ml tin of paint.

Now, the total litres of paint Jerry need to buy is 2-L + 1/2 -L = 5/2 L

Hence, the answer is 5/2 or 2 ½ Litres.

SOLUTION : Given that : Lana's illustration = 6 cm wide

Moreover, it is also given that the area of an illustration = 48 cm^2

To calculate : The enlarged illustration 's area after Lana doubles the width and the height

As clearly seen in the picture, the illustration is in the rectangular form.

Given that it is 6 cm wide, thus breadth is 6 cm.

Given: Area of illustration = 48 cm^2.

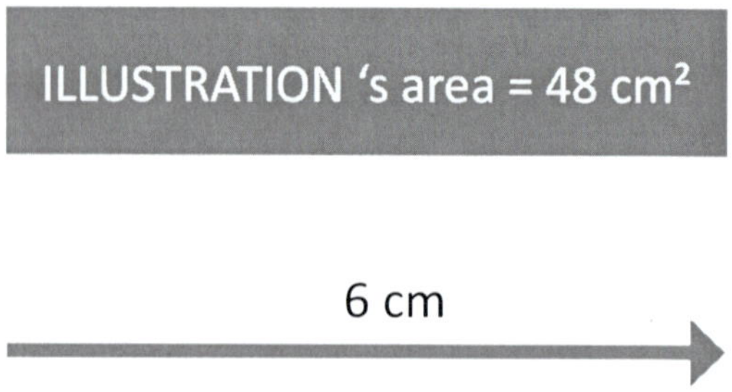

Means, Height X Breadth = 48 cm^2 = L x 6 cm = 48 cm^2

$\therefore$ H = 48 cm^2/ 6 cm = 8 cm.

Height, H = 8 cm.

The question further states that Lana wants B doubled and H doubled.

∴ B becomes 2 B and H becomes 2 H.

We have to calculate the area of the enlarged illustration, which will be:

(BREADTH OF ENLARGED ILLUSTRATION X HEIGHT OF ENLARGED ILLUSTRATION)

$$2\ B \times 2\ H = 2 \times 6\ cm \times 2 \times 8\ cm = 12\ cm \times 16\ cm = 192\ cm^2.$$

∴ The area of the enlarged illustration is 192 cm^2.

BEGINNER MEASUREMENT AND SPACE

2- D SHAPES

SOLUTION 1: To find : A 2-D shape that has 6 sides and 6 vertices

- ○ Octagon
- ○ Pentagon
- ● Hexagon
- ○ Polygon

HINT:

- An Octagon is a closed figure that has 8 sides and 8 vertices.
- Similarly, A pentagon will have 5 sides and 5 vertices.
- A hexagon always has 6 sides and 6 vertices.
- A polygon is a closed figure that has at least than 3 sides and 3 vertices.

Hence, the correct option is (C) Hexagon.

SOLUTION 2 : To find : Axes of symmetry in a rectangle

● 2 ○ 4 ○ 6 ○ 8

HINT: An Axis of Symmetry is a line that divides the object into two equal halves, thereby creating a mirror-like reflection of either side of the object.

Thus, the rectangle can be divided into two equal halves. See the figure for explanation.

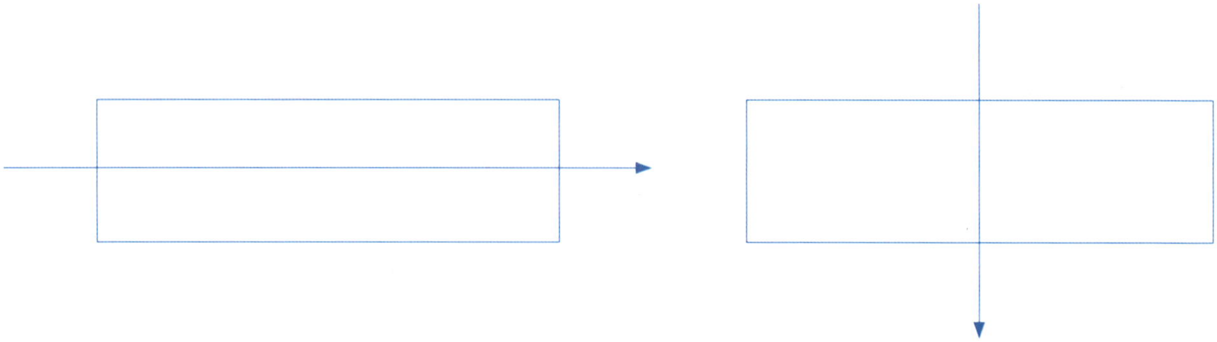

So, the option (A) is correct.

SOLUTION 3 : To find : The shape that is odd one out

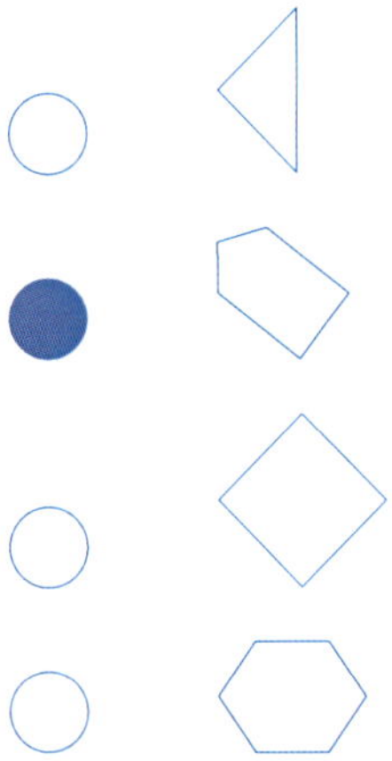

Clearly, option (B) has an irregular shape, that is an irregular pentagon.

So, option (B) is correct.

ANGLES ON A COMPASS

SOLUTION 1 : An acute angle is an angle which is less than 90° but more than 0°.

[0°<ACUTE ANGLE< 90°]

Now, as clearly seen from the figure, N is at 360° and NW is at 315°,

Hence their difference is 360°- 315° = 45°.

Therefore, Answer = 45°

HOT TIP :

LOOK AT THE GIVEN DIAGRAM OF THE COMPASS IN YOUR BOOK !!

SOLUTION 2 : An obtuse angle is always greater than 90° and less than 180°.

[90°<OBTUSE ANGLE<180°]

Now, as clearly seen from the figure, E is at 90° (Right angled triangle) and SW is at 225°.

Hence their difference is 225° - 90° = 135°.

Therefore, Answer = 135°

AREA OF TRIANGLES

NOTE THAT WATCH THE DIAGRAM CAREFULLY IN YOUR NAPLAN BOOK :

SOLUTION : Given that : In the diagram, area A + area B = 18 cm^2 & area D = 10 cm^2.

To calculate : 1) The area of Δ B

2) The area of Δ C

Let us find out the shapes of A, B, C and D respectively.

From the figure, it is clearly observed that

A = Rectangle, so area A = length x breadth

Similarly, D= Rectangle, so area D = length x breadth

Moreover, B= right angled triangle,

Thus, area B = ½ X base X height

Similarly, C = Right angled triangle

$\therefore$ area C = ½ x base x height.

Let us also assume the points a, b, c, d, e , f.

Given that: AF = 3cm= BG.

Also, given that, area A + area B = 18 cm^2

$\therefore$ length (AB) x 3 cm^2 + ½ x3 x 2 = 18 cm^2

3 AB + 3 cm^2= 18 cm^2

AB = 18 cm^2 - 3 cm^2/3 = 15 cm^2 / 3 cm = 5 cm -------------1st equation

∴ AB = 5 cm = FG = DE -------------2nd equation

Also, given that area (D) = 10 cm^2

∴ length GF x breadth DG = 10 cm^2

5 cm x DG = 10 cm^2

∴ DG= 2 cm -------------3rd equation

Using all the equations, we can conclude that,

Area of triangle B= ½ x BG x CG

∴ ½ x 3cm x 2 cm = 3 cm^2

Similarly, area of triangle C = ½ x GD x CG

∴ ½ x 2 cm x 2 cm = 2 cm^2

Hence, 1) the area of triangle B = 3 cm^2 is the correct answer.

And 2) the area of triangle C = 2 cm^2 is the correct answer.

TIMELINES AND TIME

SOLUTION : Given that : The timeline of Anton's 4-week holiday in Africa = 14 cm long

The number of significant events during holiday period = 5 ----------(Given)

To calculate : 1) 1 week in cms.

For 4 weeks, the timeline is 14 cm long.

Thus, 4 weeks = 14 cm in timeline -------------1st equation

So, we will have to find out the timeline for 1 week,

Dividing equation 1, by 4, we get

⇒ 4 weeks /4 = 14 cm of timeline / 4

∴ 1 week = 7 /2 = 3.5 cm

Hence, the timeline for Anton's 1st week is 3.5 cm.

2) Number of days between canoe trip and crocodile sightings

As per the figure, the canoe trip falls on 6th April and , Crocodile sightings fall on 22nd April.

Now, calculating the actual difference between these two sightings will be;

⇒ 22nd April – 06th April = 16 days

Therefore , the answer is 16 days.

Alternatively , calculate the number of days from the diagram, as each week is 7 days and 3.5 cm in timeline.

3) Given that : Anton saw elephants on = 16th April

To find : Did Anton see elephants before or after the cheetahs ?

From the given diagram, we conclude that cheetah drinking's event happened on 17th April.

Thus, Elephants seeing was "Just Before" the Cheetahs drinking.

SOLUTION : Given : Maria needs to be at the dentist at = 16:45

Total time taken by bus to reach to dentist = 35 minutes ------------(Given)

Note down : Buses can only leave from home on QUARTER HOUR .

To calculate : 1) Latest time Maria should catch the bus .

> **HOT TIPS:**
>
> 16:45 is the 24 – hours equivalent of 4:45 pm , on the quarter hour is at 00.15.30 and 45 minutes past each hour.

Now, let us assume possible cases If Maria

Catches the bus at 16:15, and as the bus at minimum takes 35 minutes, she will reach the dentist at 16:50 which is late by 5 minutes.

Best possible case: If Maria catches the bus at 16:00, she will reach at 16:35, before exactly 10 minutes that as of her appointment time. So, the most suitable time to reach is 1600.

Hence, the answer is 16:00.

2) After catching bus at 16:30 , the total time Maria could be late by .

When Maria catches the bus at 16:30 and adding 35 minutes to it , that is , 17:05.

But she has to be at dentist at sharp 16:45 , hence , difference between both times is 20 minutes . [⊠ 17: 05 – 16: 45 = 00: 20 *in the time format*]

So , the correct answer is 20 minutes .

BEGINNER ALGEBRA, PATTERNS AND DATA

NUMBER PATTERNS AND PROBLEMS

SOLUTION: Mai's plan for 1st week = 15 dollars ------------(Given)

To calculate : Mai 's savings plan until the eleventh month.

She saves 15 dollars each month; thus, we can increase 15 dollars every month as her savings plan to purchase clarinet.

Mai's plan until 2nd week =15 + 15 = 30 dollars

Similarly, Mai's plan until 3rd week = 30 + 15 = 45 dollars

Similarly, Mai's plan until 4th week = 45 + 15 = 60 dollars

Similarly, Mai's plan until 5th week = 60 + 15 = 75 dollars

Similarly, Mai's plan until 6th week = 75 + 15 = 90 dollars

Similarly, Mai's plan until 7th week = 90 + 15 = 105 dollars

Similarly, Mai's plan until 8th week = 105 + 15 = 120 dollars

Similarly, Mai's plan until 9th week = 120 + 15 = 135 dollars

Similarly, Mai's plan until 10th week = 135 + 15 = 150 dollars

Similarly, Mai's plan until 11th week = 150 + 15 = 165 dollars

Putting these values back into the table,

Month	**1**	**3**	**5**	**7**	**9**	**11**
$ Savings	**15**	**45**	**75**	**105**	**135**	**165**

SOLUTION : Sal gets a delivery of pallets = 6 ------------(Given)

Number of boxes in each pallet = 10 ------------(Given)

Number of boxes used by restaurant per week = 12 ------------(Given)

To calculate : Number of boxes in the storeroom at the beginning of each week.

Week	1	2	3	4	5	6
No. of boxes	60	48	36	24	12	0

Number of boxes for the first week = 60 ------------(Given)

Boxes used after 1st week = 12

So, the number of boxes left for 2nd week = 60 – 12 = 48

Similarly, for the third week = 48 – 12 = 36

Going on , for the fourth week = 36 – 12 = 24 ------------(Given)

Similarly, for the fifth week = 24 – 12 = 12

Lastly, the number of boxes left for the sixth week are ;

=> 12 – 12 = 0

SOLUTION : Given that : The Koala sanctuary has in its first year = 17 koalas

The numbers of Koalas increased (per year) = 11

To calculate : Population after every year till 5 years.

Year	1	2	3	4	5
No. of Koalas	17	28	39	50	61

Number of Koalas in the second year = 17 + 11 = 28

Number of koalas in the third year = 28 + 11 = 39

Number of koalas in the fourth year = 39 + 11 = 50

Finally, the Koalas in the fifth year = 51 + 11 = 61

NUMBER SENTENCE

SOLUTION: To Find: Number of dogs in the town.

Given that: 24 families have 1 dog each = 24 x 1 = 24 dogs

Also, Quarter (1/4) of 24 families have 2 dogs

Therefore, 1/ 4 x 24 = 6 families have 2 dogs each

We conclude that, 6 x 2 = 12 dogs

Also, 4 families own 3 dogs each = 4 x 3 = 12 dogs

Adding all the above statements, the total number of dogs in the town are: 24 + 12 + 12 = 48 dogs

In the number sentence, we can also re-write as:

$\Rightarrow$ 24 x 1 + 6 x 2 + 4 x 3 is the answer.

DATA PRESENTATION

Note down : See the table carefully in your NAPLAN book .

SOLUTION:

1. F

   (As the students' progress in school, they are less likely to spend more time with friends).

2. NI

   (There is no connection between High school students playing musical instruments and doing shopping)

3. T

   (Primary students that play Musical instruments = 32 and high school students that play these = 15)

As 32 is about twice as 15, so it true.

4. NI

   (No data provided about non-friends)

BEGINNER ALGEBRA , PATTERNS AND DATA

NUMBER PLANE

See the given number place carefully !

To plot : (3,0) , (0, -5) and (-3,0) and join these 3 points with 2 straight lines .

FUN FACT:

In (a , b) symbolizes, a for Abscissa and b for Ordinate. If the abscissa is zero , the line lies on y-axis and if the ordinate is zero , the line lies typically on x-axis.

SOLUTION 1 : The three given points are (3,0); (0, -5) and (-3,0).

When we plot these three points on the actual graph, we conclude they will make a letter V.

SOLUTION 2 : To find : The axis of points (3,0) and (-3,0) .

X – axis because the abscissa holds some value, but the ordinate is zero. It means it does not lie on y-axis but the x-axis.

SOLUTION 3 : Here, the abscissa is zero, so the given point will lie on Y-axis.

SOLUTION 4 : When both values are negative, the point always lies in the third quadrant.

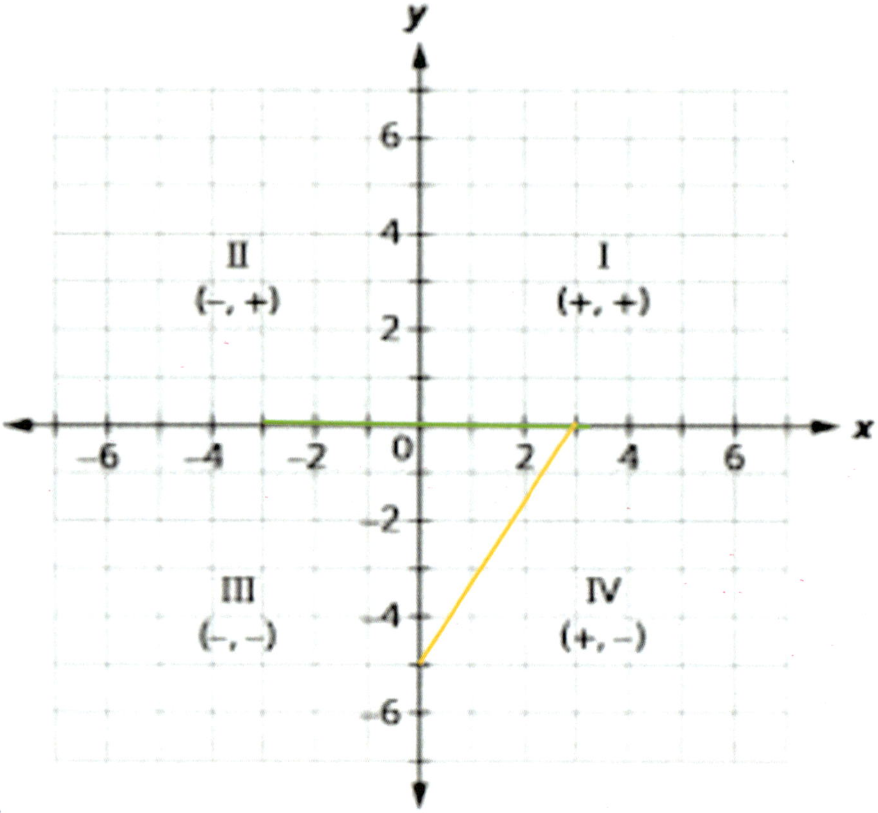

{{SEE THE ABOVE FIGURE FOR MORE CLARIFICATION}}

SOLUTION 5 : A line that joins the two given points on the number line will be parallel to x-axis.

Any line that is parallel to x-axis will be always Horizontal.

So, the answer is Horizontal.

ORDER OF OPERATIONS

(Note: BODMAS rule applies here)

SOLUTION 1 : 8 x 4 – (2 / 2) = 8 x 4 – (1) = 32 – 1 = 31

SOLUTION 2 : 15 + 3 x 2 = 15 + 6 = 21

SOLUTION 3 : 12 – 9 / 3 = 12 – 3 = 9

FUN FACT :

We solve any equation using BODMAS rule.

What it STANDS FOR :

B- Bracket solve

O- Orders of powers or root

D & M - Division and Multiplication

A & S – Addition and Subtraction

SOLUTION 4 : 36 / (4 + 2) = 36 / 6 = 6

SOLUTION 5 : 12 x 4 – 2 / 2 = 12 x 4 – 1 = 48 – 1 = 47

SOLUTION 6 : 12 x (4 – 2) / 2 = 12 x 2 / 2 = 12 x 1 = 12

SAMPLE TEST

SOLUTION 1: Solving these two mixed fractions, we get as

51/4 = 12.75 , Similarly 7/ 25 = 0.28

Thus, option B is the correct answer.

SOLUTION 2: Lachlan has dollars to spend on Christmas gifts = 120

Lachlan wishes to spend dollars on his parents = 40

Rest, money goes equally to his 5 friends (as per question)

Hence, we need to find the remaining amount = 120 – 40 = 80 dollars

Therefore, Lachlan will spend this amount of money on his 5 friends = 80/5 = 16 dollars

Hence, Option 1 is correct.

SOLUTION 3: Length of sheet = 5 m

Also, Breadth of sheet = 2 m

Here, Area of Banner that Samuel wants = Total area of rectangular sheet – 4 * area of triangles {there are four corners so 4 triangles will be cut down }

Area of triangle = ½ x base of triangle x height of triangle

Therefore, total area of Banner that Samuel wants =

Length of rectangle * breadth of rectangle – 4 * ½ * base * height

= 5 m * 2 m – 4 * ½* 1 * 1 = 10 m^2 – 2 m^2 = 8 m^2

SOLUTION 4: As clearly seen from the option 2, 8 + 3 / 1 + 3 = 11 / 4 which is not equal to 8.

SOLUTION 5: As seen clearly from the diagram, the dotted line would meet at 4.

The y- coordinate of the new point whose x- coordinate is 2 will be 4.

SOLUTION 6: The table displays the temperature where Sima lives in the United States:

Fahrenheit scale	40	60	80
Centigrade scale	4.4	15.5	26.6

We must find for 20^0 F in Centigrade,

It is known from the figure that as the temperature increases in F by 20, simultaneously the temperature in C increases by 11.1 .

Therefore, it is obvious that to find for 20^0 F, we reduce 11.1 from 4.4

Thus, the result is 4.4 – 11.1 = - 6.7^0 C

SOLUTION 7: 13 + 47 * 3 will be equal to 13 + 141 = 154

SOLUTION 8: To solve: 3 km plus 50

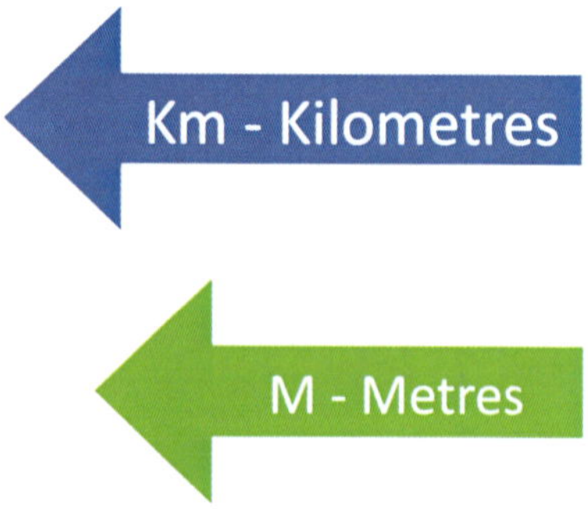

Note = 1 km = 1000 m

Therefore, 3km= 3 * 1000 m = 3000 m

Add 50 m to it, 3000 + 50 = 3050 m

Clearly, option A and D don't match,

Divide by 1000 to convert it into km,

It gives us, 3050 / 1000 = 3.05 km

Thus, Option 3 is the correct answer.

SOLUTION 9 : 21041 / 6 =

			3	5	0	6			
		6	2	1	0	4	1		
			1	8					
			-	3	0				
				3	0				
				-	0	4	1		
					-	3	6		
							5		

From the DIVISION shown in the table added up there,

The remainder of the question is 5, the dividend is 21041, the divisor is 6 and the quotient is 3506 !

To write it in mixed fraction, $3506 \frac{5}{6}$ is the correct answer.

SOLUTION 10: Reflex angle is an angle that is more than 180 degrees and less than 360 degrees.

In this case, option A is right-angled triangle.

Option B is also a right-angled triangle.

Option C is an angle of 135^0 so it is not a reflex angle.

Option D is an angle that is more than 180^0 and less than 360^0.

Hence, Option D is correct.

SOLUTION 11: Sienna has spent on drinks = 12 dollars

Then, she purchased food for dollars = 10

To find : How much money should Sienna get from 4 of her friends

Note : If Sienna spends on her 4 friends and including her would be 5, so 12 dollars / 5 persons = 2.4 dollars / person

Thus, each is paying 2.4 dollars to share the drinks equally.

Therefore , Sienna needs money from her FOUR friends would be
4 * 2.4 = 9.60 dollars

So, 9.60 dollars is the correct answer

SOLUTION 12 : Given Numbers : Sum Of Digits :

Given Numbers	Sum Of Digits
465	15
1092	12
63	9
7425	18
95412	21

Note : These five numbers could be divisible by a whole number , so that the answer could be a whole number as well.

FUN FACT: A whole number is a number without fractions/ decimals.

$\therefore$ It is an integer.

{0,1,2,3,4....}

As it is clearly seen that the sum of digits of above numbers are divisible by 3 which is also a whole number.

Thus, 3 option B is the correct choice.

SOLUTION 13.

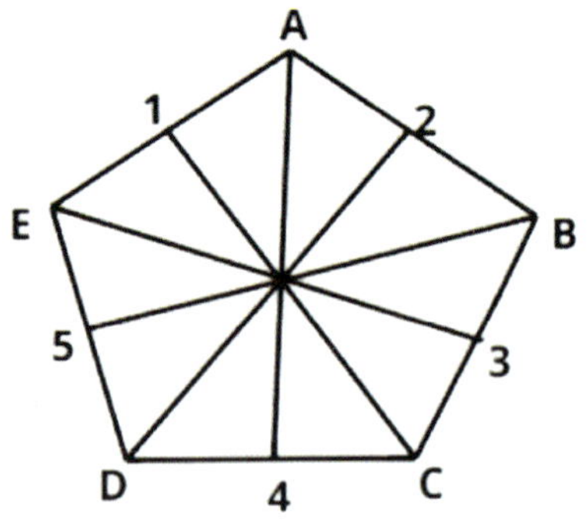

DID YOU KNOW :

An axis of symmetry : A line through a shape so that each side is a MIRROR IMAGE. Note that a REGULAR PENTAGON has 5 sides and 5 lines of Symmetry. In this case, the number of lines of Symmetry in a regular pentagon is equal to the number of sides.

Point to remember : Polygon is a figure , e.g., Pentagon.

Learn that : Each line of Symmetry goes from a vertex to the mid-point of opposite side.

Therefore, the Axes of Symmetry for a REGULAR PENTAGON is 5 making Option A as the correct option.

SOLUTION 14 : 355,750 TO THE NEAREST 10,000 , according to the place value system table would be 360,000 because the number would be closer to 60,000 .

Hence, the correct answer is 360,000 .

SOLUTION 15 : Let us observe the following pattern carefully:

0.75 4.6 8.45 12.3

To find: Sixth number in the Pattern.

Sol: Let us try to figure out the difference of these consecutive digits, 4.6 - 0.75 = 3.85

The above pattern MUST follow this difference.

Now, add 3.85 to 4.6 which gives us exactly 8.45 and so on.

Finally, the fifth number will be 12.3 + 3.85 = 16.15

Similarly, to calculate the sixth number of given sequence = 16.15 + 3.85 = 20.00

Thus, 20 is the right answer.

SOLUTION 17. Balmoral hospital Raises dollars = 85000

Total teams of Volunteers = 19

∴ Average raised by each team = (85000 / 19) dollars = 4473.68 dollars

As it is quite close to $ 4500.

So, option (B) is the correct answer.

SOLUTION 18. As the given figure is, a HEXAGON with 6 straight lines , it must have 6 interior angles that each measure is 120^0, sum to 720^0.

Why ? As we know that , $\angle b + \angle\frac{c}{2} + \angle o = \angle 180$ ......equation 1 {By triangle sum property} [o is the angle at the center]

Also, it is an Equilateral triangle as all 3 sides are equal.

$\therefore\ \angle b = \angle\frac{c}{2} = \angle o$ ......equation 2

Put equation 2 inn equation 1 , we get

Hence, $\angle\frac{c}{2} + \angle\frac{c}{2} + \angle\frac{c}{2} == \angle 180$

$3\ \angle\frac{c}{2} = 180\ \therefore\ \angle c = 60\ X\ 2 = \angle 120$

Therefore, The angle at the vertex of this regular polygon is 120^0, making option C as the correct choice.

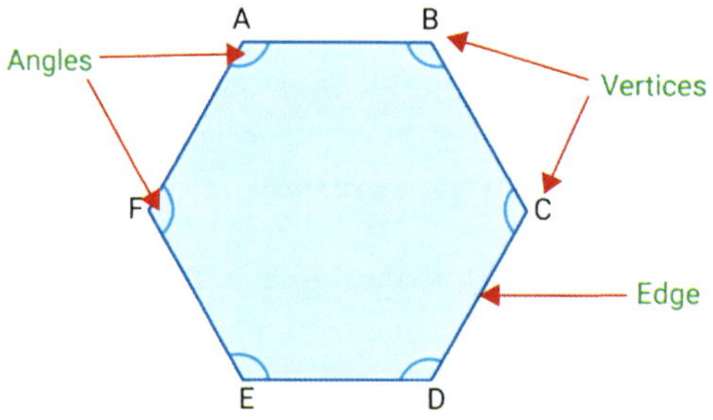

Hexagon

SOLUTION 19. Results for maths test at Mayberry High is shown in the following table :

Score	**% Students**
41-50	7
51-60	22
61-70	45
71-80	15
81-90	8
91-100	3

To calculate : Percentage of students scored between 51 and 80 in the test.

Clearly, between 51 & 80 , it could be 3 possible ways :

22 % , 45% , 15%.

To add, we get 82 % .

∴ 82 % is the correct answer.

SOLUTION 20. Given is the octagonal pyramid shape , (Combination of 8 triangles) .

Base of each triangle = 3 metres (given)

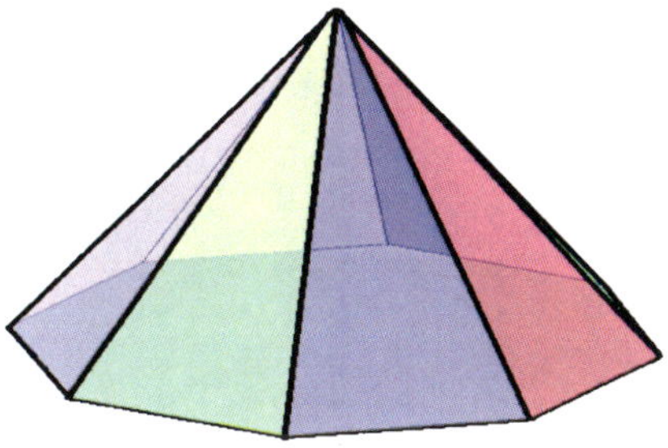

Perpendicular of triangle = 12 metres (given)

Thus, total area of triangular surfaces = 8 X Area of each triangle = 8 X ½ X base of Triangle X height of Triangle

= 8 X ½ X 3 m X 12 m = 144 m^2 .

Hence, 144 m^2 is the total area of the triangular surfaces.

SOLUTION 21. City's population (before 2 years) = 2560000

Rise in City's population for 1st year = 45000

Rise in City's population for 2nd year = 45000

∴ Total rise in population in 2 years combined = 45000 + 45000 = 90000

∴ Current population = 2560000 + 90000 = 2650000

It is certainly equal to 2.65 Millions. [1 Million = 10,00,000]

So, option 3 is the correct option.

SOLUTION 22. Perimeter of Park = 340 m = Sum of all sides of Park

Breadth = 60 m (given)

Let us assume that length of park is a meters,

∴ Perimeter = (a + a + 60 + 60) m =(2a + 120) m

Also, perimeter = 340 meters (given)

∴ 2a + 120 = 340 m

∴ 2a = 220 m

∴ a = 110 m

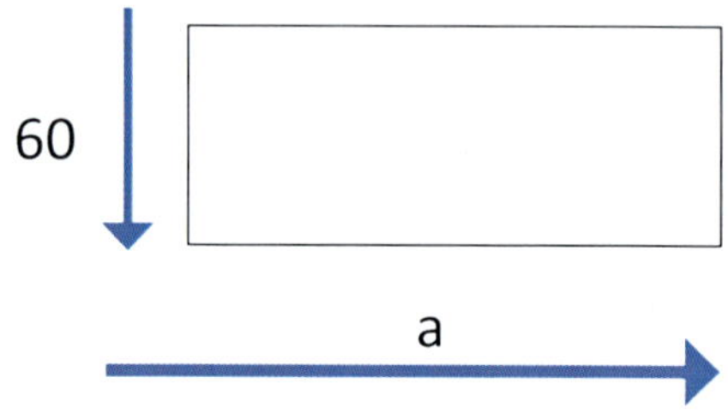

Therefore, the length of the park is 110 m.

SOLUTION 23. Given that : Time in Perth = 2 hours behind the time in Brisbane

Lou calls her friend Jinny in Brisbane = 16 : 20 (given)

∴ 16 : 20 in Perth time = 18 : 20

∴ After 10 minutes phonically conversation, Brisbane time would be (18:20 + 10) minutes => 18 : 30

Thus, option A is the correct one.

MEDIUM NUMBER

MONEY

SOLUTION : Total money that Jayden has = 50 dollars

Jayden buys = 2 loaves of breads + 2 bottles of milk + 1 carton of eggs + 9 skinny sausages

∴ Jayden spends = 2 x 4.60 + 2 x 2.50 + 3.40 + 6.60 = 24.2 dollars

∴ Jayden receives a change of = (50 – 24.2) $ = 25.8 $

Hence, 25.8 dollars is the correct answer.

FUN FACT : 6 sausages = $ 4.40

∴ 3 sausages = 4.40 / 2 = 2.20 dollars

∴ 9 sausages= 4.40 + 2.20 = 6.60 dollars

SOLUTION : Seb has got in e-book account = 405.12 dollars

∴ Cost of downloading 1 e-book = 9.85 dollars

∴ As Seb downloads 11 books ,cost of all these downloads would be equal to (11 X 9.85) $= 108.35 $

∴ After downloads, Money left in Seb's account is

405.12 $ - 108.35 $ = 296.77 dollars

Thus, 296.77 is the amount left after all eleven downloads.

SOLUTION : Sandy needs = A fishing bag + A new reel

Bag's original price = 42 dollars (given)

Discount of 20 % of bag = 42 X 20/100 = 8.4 $

∴ New discounted price of bag = 42 $ - 8.4 dollars = 33.6 dollars

Price of reel = $ 53.80

∴ According to the question, if Sandy buys a combo pack, she will pay $ 95.

So, new price of bag + price of reel + price of gloves is all together 33.60 $ + 53.80 $ + Price of gloves = 95 $

∴ 87.40 $ + Price of gloves = 95 $

∴ Price of gloves = $ 95 - $ 87.40 = $ 7.60

∴ The Price of gloves is $ 7.60.

SOLUTION :

1. 7 DVD's price = (7.95 X 7) $ = $ 55.65

   After discount of 25 % , Aya saves = 55.65 X 25/100 = 13.91 dollars

   Thus, $(55.65 – 13.91) = $ 41.74 rounded up to nearest 5 cents as per the question.

   ∴ $ 41.75 is the amount that Aya pays.

2. Ben made a profit = $ 177.92

   According to the question, this profit must be shared between his 2 brothers and himself.

   So, it will be shared among 3 people.

   ∴ Each person will receive = 177.92/3 = 59.306

   Thus, Rounded up to nearest 5 cents , it becomes 59.30 $.

   Hence, 59.30 dollars is the correct answer.

SPECIAL NUMBERS

SOLUTION : To find : 6 odd numbers that must be satisfying....

- Greater than 2 and less than 50
- Neither prime nor 5 as a factor

  Let us begin by finding prime numbers ; (between 2 and 50)

FUN FACT : The numbers that are not PRIME are CALLED COMPOSITE numbers and VICE-VERSA !!!!!!!

∴ 3, 5, 7, 11, 13, 17, 19, 23 , 29 , 31 , 37 , 41 , 43 , 47

If these are prime numbers, then non- prime number would be odd but simultaneously satisfy above two DOT points, we obtain

9, 15, 21, 25 , 27, 33, 35, 39 , 49 => Composite numbers

Now, moving on and excluding numbers that have 5 as a factor

∴ 6 numbers are 9, 21 , 27 , 33, 39 ,49 which is the CORRECT answer.

SOLUTION : 1. 36 = 6 X 6 , so , the number is 6.

2. 144 = 12 X 12 , so , the number is 12.

3. 81 = 9 X 9 , so , the number is 9.

4. 16 = 4 X 4 , so , the number is 4.

SOLUTION : 1. $3^3 = 3 \times 3 \times 3 = 27$

2. $5^3 = 5 \times 5 \times 5 = 125$

3. $4^3 = 4 \times 4 \times 4 = 64$

4. $2^3 = 2 \times 2 \times 2 = 8$

SOLUTION :

1. From the given sequence in the Question, 3 6 10 15 21 28 36.... this category of numbers FORM a TRIANGLE of Circles ; thus, it is called "TRIANGULAR" ,

X	X	X	X	X
O	X	X	X	X
O	O	X	X	X
O	O	O	X	X
O	O	O	O	X

2. AS it is clearly seen from the given sequence of numbers, the third number is 10 and the fourth number is 15.

   To find : A number sequence that includes both third and fourth numbers , equals $5^2 = 25$

   $\therefore$ $10 + 15 = 25 = 5^2$ is the correct answer.

MEDIUM NUMBER

FRACTIONS AND DECIMALS

1. 16 / 28 = 8 / 14 = 4 / 7
2. 45 / 100 = 9 / 20
3. 27/ 63 = 3 / 7
4. 20 / 50 = 2 / 5

1. 15 / 4 = 3 ¾
2. 200 / 15 = 40 / 3 = 13 $\frac{1}{3}$
3. 183 / 12 = 61 / 4 = 15 ¼
4. 85 / 35 = 17 / 7 = 2 $\frac{3}{7}$

1. 1/4 + 3/8 = 2/8 + 3/ 8= 5/8
2. 1 $\frac{2}{3}$ + 5/8 = 5/3 + 5/8 = 5/3 X 8/8 + 5/8 X 3/3 = 40/12 + 15/24 = 55/ 24
3. 8/15 – 1/6 = 8/15 X 6/6 – 1/6 X 15/15 = 48/90 – 15/90 = 33/90 = 11/30
4. 23/15 – 2/3 = 23 – 5 X 2 / 15 = 23 – 10 / 15 = 13 / 15 (Note that here 15 is the L.C.M. itself as 3 is a factor 15 & L.C.M. is known as Lowest Common Factor)
5. 3.52 + 0.90 = 4.42
6. 12.70 – 8.63 = 4.07
7. 16.00 – 8.54 = 7.46
8. 3.60 + 0.78 = 4.38
9. 7.06 – 4.29 = 2.77
10. 6 $\frac{3}{8}$ + 2 $\frac{5}{8}$ = 51 / 8 + 21 / 8 = 72 / 8 = 9

> FUN FACT :
>
> Make both denominators EQUAL by multiplying and dividing by the same number. This number would be the first and second denominator of the both FRACTIONS !!

> HINT:
>
> Solve Mixed Fraction first !!

1. $0.37 \text{ X } 100 = \frac{37}{100} \text{ X } 100 = 37$
2. $\frac{2}{3} \text{X} \frac{1}{8} = \frac{2}{24} = \frac{1}{12}$

1. $98.5 \div 10 = \frac{985}{10} \text{ X } \frac{1}{10} = \frac{985}{100} = 9.85$
2. $\frac{3}{8} \div 2\frac{1}{4} = \frac{3}{8} \div \frac{9}{4} = \frac{3}{8} \text{ X } \frac{4}{9} = \frac{12}{72} = \frac{1}{6}$ [Reciprocal of $\frac{9}{4}$ is $\frac{4}{9}$]
3. $0.303 \text{ X } 9 = \frac{303}{1000} \text{ X } 9 = \frac{2727}{1000} = 2.727$
4. $20.52 \div 0.02 = \frac{2052}{100} \div \frac{2}{100} = \frac{2052}{100} \text{ X } \frac{100}{2} = \frac{2052}{2} = 1026$
5. $3\frac{1}{3} \div \frac{2}{3} = \frac{10}{3} \text{ X } \frac{3}{2} = 5$
6. $4\frac{1}{2} \div 6 = \frac{9}{2} \text{ X } \frac{1}{6} = \frac{3}{4}$
7. $4.5 \div 0.45 = \frac{45}{10} \div \frac{45}{100} = \frac{45}{10}\ X\ \frac{100}{45} = 10$
8. $1\frac{1}{5} \text{ X } \frac{5}{9} = \frac{6}{5} \text{ X } \frac{5}{9} = \frac{6}{9} = \frac{2}{3}$

POINT TO BE NOTED :

When we change the sign (÷) to sign (X) in any two fractions that are MULTIPLIED by each other, the second fraction automatically becomes reciprocal of it.

MEDIUM MEASUREMENT AND SPACE

ANGLES INSIDE THE TRIANGLE

SOLUTION : From the figure given in your NAPLAN workbook ,

h = 119 [Vertically Angles Theorem]

Also, using same theorem , g = i ,

$$g + i + 119 + 119 = 360 \text{ [Angle Sum Property]}$$

$$\therefore 2g = 360 - 119 - 119 \Rightarrow 122$$

$$\therefore g = 61 = i$$

Also, $a + h + 36 = 180$ [Triangle Sum Property]

$$a + h = 180 - 36$$

$$\therefore a + 119 = 144$$

Thus , $a = 25$ .... 1st equation

Given that : $a + b = 60$, by putting equation 1 , we get

=> $b = 35$

Again, by Triangle Sum Property applying in Δ BCG,

=> $b + c + g = 180$ [$\because g = 61$]

=> $35 + c + 61 = 180$

=> $c = 180 - 96 = 84$

Similarly , again using the Triangle Sum Property ,

=> $36 + i + f = 180$

=> $36 + 61 + f = 180$ [$\because i = 61$]

=> $f = 180 - 97 = 83$

As $f + e = 116$ .... (Given)

=> $83 + e = 116$

=> $e = 33$

Similarly , again using the Triangle Sum Property ,

=> $e + d + 119 = 180$

=> $33 + d = 180 - 119$

=> $33 + d = 61$

Hence, $d = 61 - 33 = 28$

Therefore, the value of each labelled angle are as follows:

1. a = 25 ; 6. f = 83
2. b = 35 ; 7. g = 61
3. c = 84 ; 8. h = 119
4. d = 28 ; 9. i = 61
5. e = 33

SOLUTION 10. angles g and h => Adjacent angles

SOLUTION 11. angles 119^0 and h => Vertically opposite angles

SOLUTION 12. angle g is $\angle 61$; so it is an ACUTE angle.

SOLUTION 13. angle h is $\angle 119$; so it is an OBTUSE angle.

> HOT TIP :
>
> Sum of ADJACENT angles is always 180^0.
>
> VERTICALLY OPPOSITE angles are always EQUAL.
>
> An ACUTE angle ranges between 0^0 and 90^0.
>
> An OBTUSE angle ranges between 90^0 and 180^0.

VOLUME MEASUREMENTS

SOLUTION 1 : Initially, the water is 1 litre in a 2 liters cylinder.

Put the smaller wooden piece ⟷ 0.723 litre water level rise .... (Given)

To find : Volume of smaller piece of marble in cm^3

As we know that by putting the stone,

=> Volume of the stone = Volume of the marble

∴ Volume of marble = 0.723 litre

But we need to convert it into cm^3 as per the QUESTION ,

Also, 1 ml = 1 cm^3 ....1st equation (ml - milliliter)

& 1 litre = 1000 ml

∴ 1 L = 1000 ml X 1

So, 0.723 L = 1000 ml X 0.723 = 723 ml

Using equation 1 , 723 ml = 723 cm^3

Hence, the answer is 723 cm^3 .

REMEMBER THAT:

1cm^3 = 1 ml

1 m^3 = 1000 litres

1 litre = 1000 ml

FUN FACT :

1 cm = 10 mm

1 cm^2 = 10 mm X 10 mm = 100 mm^2

Similarly ,

1 cm^3 = 1000 mm^3

SOLUTION 2 : Volume of larger stone = length l X breadth b X height h

∴ Volume of larger stone = 110 mm X 140 mm X 190 mm => 2926000 mm^3

But , according to the question, we require to find the volume of Marble stone in cm^3,

Therefore , we will convert 2926000 mm^3 into cm^3.

As, 1 cm^3 = 1000 mm^3

1 cm^3 = 1000 mm^3 X 1

∴ 1 mm^3 = $\frac{1}{1000}$ cm^3

∴ 2926000 mm^3 = 2926000/1000 cm^3

Volume of Marble stone in cm^3 = 2926 cm^3 is the correct answer.

REMEMBER THAT :

Marble is of cuboid shape as $l \neq b \neq h$

MEDIUM MEASUREMENT AND SPACE

MAPS AND GRIDS

SOLUTION 1 : After seeing the Map in your NAPLAN workbook,

E2 > Emily's house

B3 > Drew's house

And B2 > Post office

{***When Emily headed NORTH for 6 Kms from Drew's House***}

∴ From B2 to E2, the distance is 18 kms (9 cms) < as 1 cm = 2 km (Given)>

> The direction is East.

∴ Emily headed to home from post office at east for 18 kms.

MASS

SOLUTION : See the weighting scale carefully, the arrows are (approximately) 2 kgs, 40 kgs , 80 kgs, 450 kgs respectively.

After looking at the weights, we conclude that ,

Large dog = 40 kgs means Arrow 2.

Polar beer = 450 kgs means Arrow 4.

2 L bottle of milk = 2 kgs means Arrow 1.

Adult person = 80 kgs means Arrow 3.

SPEED

SOLUTION : Distance swam by Josh = 200 m

Time taken = 3 min , 10 seconds = 190 seconds

To Find : Average speed (up to 2 decimal places)

As per the Speed – Distance formula,

Time = Distance / Average Speed

∴ Average speed = Distance / time = 200 m / 190 sec

= 20/19 m/s or 1.05 m/sec

NOTE DOWN :

Preferrable units of time with km is minutes and with metres is seconds.

∴ We converted minutes into seconds.

SOLUTION : Given that : One lap = 4 kms long

As Frankie is 6 laps from the race finishes= 6 X 4 = 24 kms

Here, Distance = 24 kms

Speed = 180 km/hr.

To Find : Time = ?

Again, as per the speed – distance formula , we have

Time = Distance / Speed = 24 km / 180 km/h = 2/15 h

∴ 2/15 X 60 minutes = 8 minutes

Therefore, 8 minutes is the correct answer.

Convert into minutes !!

As 1 hour = 60 minutes !!

SOLUTION :

i. Legal speed limit in a School zone= 40 km/ hr

But we require to convert km / hr into m /s .

∴ Putting the right-hand side HINT, we obtain

40 km/ 1h = 40000 m/ 3600 sec = 400/ 36 (m/ sec)

= 11.11 m / sec is the CORRECT ANSWER .

ii. Now, when the driver is approaching the school,

His speed is legal .

∴ Speed = 11.11 m / sec

The time he used to glance at watch = 1.5 sec

As we need to calculate the distance, we use

HINT:

1 km = 1000 m

∴40 km = 40000 m

More or less,

1 hr = 60 min or 360 seconds (60 X 60)

Distance = Speed X time

D = 11.11 X 1.5 = 16. 65 metres

Hence, 16.65 metres is the correct answer.

ENLARGEMENT AND PERSPECTIVE

SOLUTION 1 & 2 : Note down : This figure is getting enlarger.

As the base of larger triangle is double than the base of smaller triangle , therefore the figure is getting enlarged by double.

So, a = 2 (5) = 10 is the correct answer.

1. If a = 10 ; Using Pythagoras theorem , we get

$(15)^2 = (12)^2 + b^2$

$\therefore$ 225 – 144 = b^2

$\therefore$ 81 = b^2

$\therefore$ b = 9 is the correct answer.

SOLUTION 3 : Enlargement factor Definition: It is the scale factor that describes the size of an enlargement.

As Perpendicular of the larger Δ is 3 times than the perpendicular of smaller Δ, thus making enlargement factor as 3.

Thus, Enlargement factor = 3

SOLUTION : Clearly from the picture, it is depicted that boat a < boat b.

Remember that boat b is far and still looks bigger whereas boat a is nearer and still looks smaller. We can , therefore , prove that the boat b is LARGER in real life.

MEDIUM ALGEBRA , PATTERNS AND DATA

FREQUENCY AND PROBABILITY

SOLUTION 1 : Frequency is the number of occurrences of a repeating event per unit of time.

∴ Calculate the total numbers of individuals that belong to certain score group.

Score group	A	B	C	D	E	F	G	H
Frequency	1	3	4	2	2	1	2	1

SOLUTION 2 : Probability is the possibility of an EVENT occurring == Definition of Probability !!

Now, the game score between 70 and 79 = 4

And total outcomes / events = 16

∴ According to formula,

$$\text{Probability} = \frac{\text{particular outcome for an event}}{\textit{total number of outcomes}}$$

$= \frac{4}{16} = \frac{1}{4}$ is the correct answer.

SOLUTION 3 : Similarly, after seeing carefully , we obtain

P (Event A, B, C, D) = P(A)+P(B)+P(C)+P(D) / No. of outcomes

Therefore, $P = \frac{1+3+4+2}{16} = \frac{10}{16} = \frac{5}{8}$

Hence, $\frac{5}{8}$ is the correct answer.

SOLUTION : Karl must choose from 1 to 15.

$\therefore$ Total number of events in this case = 1, 2, 3, 4, 5, 6, 7, 8, 9, 10, 11, 12, 13, 14, 15

Probability (10, 12 , odd number) = $\frac{10}{15}$ = $\frac{2}{3}$ [Remove 10, 12 and all even numbers]

Thus, $\frac{2}{3}$ is the correct answer.

SOLUTION : Number of boys in Rick's class = 18

No. of girls in Rick's class = 14

$\therefore$ Total number of outcomes (events) = 18 + 14 = 32

Probability of having a girl = $\frac{14}{32}$ = $\frac{7}{16}$

Similarly, probability of having a boy = $\frac{18}{32}$ = $\frac{9}{16}$

Going further, , if a teacher chooses one girl but she does not rejoin, it means she is gone.

Total number of students left now = 31

$\therefore$ Probability of remaining girls in the class = $\frac{13}{31}$

And probability of total number of boys = $\frac{18}{31}$

So, the probability of selecting a boy when a girl lefts the class would be $\frac{18}{31}$.

FUN FACT :

Probability is always between 0 and 1. It is zero when an event cannot happen whereas it is 1 when an event is surely going to happen.

FUN FACT :

Probability is also called POSSIBILITY of having an event.

RATIOS

SOLUTION 1 : Given that, number of students at Baytown High are members of an activity group or a sporting team = 294

$\therefore$ Also, given that; $\frac{Total\ Number\ of\ members}{Total\ number\ of\ students} = \frac{3}{8}$ or 3 : 8 ...1st eq"

Let the total number of students be x

Using 1st equation, $\frac{294}{x} = \frac{3}{8}$ => x = 784

Thus, the total number of students at Baytown High is 784.

1. Students (not participated) = Total number of students – students (that participated)

   $\therefore$ Students = 784 – 294 = 490

   Now, to calculate percentage,

   Percentage = $\frac{490}{784} \times 100$ = 62.5

   $\therefore$ Percentage of such students is 62.5 .

POINT TO REMEMBER :

Before finding the percentage of the students that did not participate in either an activity group or sporting team , we do require to find such number of students.

HOT POINT : % percentage formula = $\frac{value}{total\ value}$X 100

NUMBER PATTERNS

SOLUTION : To find : Estimate of members at the end of 4 years

As the club already has 20 people and each year Tara allows a 50 % increase, simultaneously a loss of 2 members per year. .... (given)

For first year, 20 + 50 % of 20 – 2 = 20 + 10 – 2 = 28

Similarly, second year = 28 + 50 % of 28 – 2 = 40

Similarly, third year = 40 + 50 % of 40 – 2 = 58

Likewise , for forth and last year = 58 + 50 % of 58 – 2 = 85

∴ At the end of 4 years , Tara would have 85 members in her Gymnastics club.

NUMBER SENTENCES

SOLUTION : Total number of people that need chair – lifts = Skiers + Instructors = 22 + 4 = 26

Given that ; as on each chair – lifts , only 4 individuals could sit, therefore, the number sentence to show the number of chairlifts are required would be (22 + 4) / 4 .

So, $\frac{22+4}{4}$ is the correct answer.

POINT TO REMEMBER :

Let us try to find the cost of candles and candle-holders separately and add them up.

SOLUTION : Cost of 100 candles = $ 145

∴ Cost of 1 candle = $ 145 / 100

As we need 180 candles, so cost of 180 candles will be

$ $\frac{145}{100}$ X 180 .... 1st equation

Similarly, the cost of 30 candle holders = $ 25

∴ Cost of 1 candle holder = $ 25/30

But we require $\frac{180}{2}$ candle holders as each candle holders could hold 2 candles at maximum. ....(Given)

$\therefore$ So, the cost of 180/2 candle holders = $\frac{25}{30}$ X $\frac{180}{2}$ .....2nd equation

Therefore, total cost of candles and candle-holders is :

$\frac{145}{100}$ X 180 + $\frac{25}{30}$ X $\frac{180}{2}$ is the Number sentence of total cost.

MEDIUM ALGEBRA , PATTERNS AND DATA

NOTE: LOOK AT THE BAR CHART CLEARLY .

SOLUTION 1 : Yes, as the bar chart clearly displays the % of individuals that are interested in Netball court or Skateboard ramp.

It also includes information about interested groups.

SOLUTION 2 : No, as we need to keep the things all together, increasingly any particular % would affect the WHOLE BAR CHART.

SOLUTION 3 : No, if the number of people who replied and not on the graph, won't be counted.

AVERAGES

Given that maximum wind speed (km/h) are as follows:

SOLUTION 1. 5.7 22.1 27.5 19.3 17.9 at North beach.

Average maximum wind speed = $\frac{15.7+22.1+27.5+19.3+17.9}{5}$

=> $(\frac{102.5}{5})$ km/h = 20.5 km/h

So, the correct answer is 20.5 km/h.

SOLUTION 2. Sixth day speed of wind = 3.7 km/h

=> $\frac{15.7+22.1+27.5+19.3+17.9+3.7}{6}$ = $(\frac{106.2}{656})$ =) km/ h = 17.7 km/h

Thus , the correct answer is 17.7 km/h.

SAMPLE TEST

SOLUTION 1. Clearly, 84 X 0.28 is first 3 options but it is NEVER equal to 0.

Look at option d 84 X 0 X 0.28 which equals ZERO.

∴ Option D is correct .

SOLUTION 2 : Highest factor or highest common factor, also known as H.C.F. in short form.

3	1251
x	417

Here, 1251 = 417 X 3

As per the question, we get 1251 when highest factor of 1251 is multiplied by a prime number.

Here, 3 is the prime number.

∴ 417 is H.C.F., which is being multiplied by the prime number to produce a result of 1251.

SOLUTION 3 : Using adjacent angle theory,

$\angle a + \angle 110 = \angle 180$

So, $\angle a = \angle 70$

Also, if the angles are

Equal, then $\angle b + \angle b$

FUN FACT :

As the triangle in the figure is an Isosceles triangle, because two sides are equal. (Given)... 1st equation

∴ In an Isosceles Δ, two angles are also equal. Following 1st equation !

$\therefore\ \angle b + \angle b + \angle a = \angle 180$ ....Δ sum property

Thus, $2\angle b = \angle 180 - \angle 70$

$\therefore\ \angle b = \angle 110 / 2 = \angle 55$

Thus, making option A as the right choice.

REMEMBER :

An angle opposite to a that lies in the Δ is equal to 70^0 because both are VERTICALLY OPPOSITE ANGLES.

SOLUTION 4 : To find : 25.27 – 8.309 rounded to 2 decimal places.

25. 270 – 8.309 = 16.961 is up to 3 decimal places.

Now, to make it to 2 decimal places, we will write 16.96.

Thus, 16.96 is the correct answer.

SOLUTION 5 : Erol is willing to fill up the pool = 6000 L

Flow rate by which the pool could be filled up =20 L/m

$\therefore$ Time taken to fill up the pool = $\frac{Quantity\ of\ pool}{Flow\ rate}$

=> 6000 L / 20 L/min = 300 minutes = 5 hrs.

Thus, option D is the correct option.

SOLUTION 6 : Given : Measurements of base of Charlotte's room = 80 mm X 80 mm

=> Initial height = 200 mm

To find : mL of water to make Charlotte's room ¾ full.

$\therefore$ Final height = ¾ X actual height

=> ¾ X 200 mm = 150 mm

Now, volume of the water room

that room has been filled with = length X breadth X height

FUN FACT :

We require to find the final height because we just need the room to ¾ of total, so the height also has to be $3/4^{th}$ of the actual height .

To put in simpler words, we need VOLUME OF WATER AND NOT VOLUME OF ROOM.

=> 80 mm X 80 mm X 150 mm = 960000 mm^3

Now, we will convert it into ml.

We know that 1 cm = 10 mm

Hence, $1cm^3$ = 10 X 10 X 10 mm^3 = 1000 mm^3

∴ 1 cm carries a capacity of 1 ml of water.

Converting 960000 mm^3 into cm^3 would give us 960 cm^3.

As 960 cm^3 = 960 ml [1 cm^3 = 1 ml]

So therefore, 960 ml is the correct answer.

SOLUTION 7 : 2 + 0.3 + 0.14 + 0.135 which could be rewritten as 2.000 + 0.300 + 0.140 + 0.135

>==> 2.575

∴ Option A is the correct answer.

SOLUTION 8 : As probability (chance) = $\frac{Likelihood\ of\ any\ particular\ event}{Total\ number\ of\ events}$

Likelihood of 2 or 1 = 5 times , total number of events= 8

∴ Probability (2 or 1 after spinning) = $\frac{5}{8}$

Thus, $\frac{5}{8}$ is the correct answer.

SOLUTION 9 : Cost of 1 apple bought by Chloe = 30 cents

Cost of 8 apples =(8 X 30) cents= 240 cents = $ 2.4

Similarly, 2 packets of soap = $ 2.40

Thus, 4 packets of soap = 2 X $ 2.40 = $ 4.80

Now, add both costs, the total money Chloe paid is $ 4.80 + $ 2.40 = $ 7.20

Thus, option C is correct.

SOLUTION 10 : Starting from Row 47, the second aisle would be the one between G & H. walking down and seeking for the 7th row, it would be 53 and turning right would give us 53 G & 53 F.

Thus, option C 53 G & F is the correct option.

SOLUTION 11 : Clearly, C option is 12 X 12 X 12, it is certainly not equal to 12 X 12^2 .

SOLUTION 12 : Rajam ran = 1.5 km of distance (given)

Time taken by Rajam = 4 min. 25 sec. = 265 sec

Average speed (in m/s) = $\frac{distance\ travelled\ in\ metres}{total\ time\ taken\ in\ seconds}$

$\therefore$ Average speed = $\frac{1500\ m}{265\ sec}$ = 5.66 m/ sec is the answer.

SOLUTION 13 : Given : Chen bought meat from the supermarket = 622 grams

Also, price of 1 kg of meat = $ 24. 50

To calculate : The money he paid for 622 grams of the meat

=> Now, as we know that 1000 gm of meat cost $ 24.50

$\therefore$ 1 gm of meat would cost = 24. 50 / 1000 $

Similarly, 622 gm of meat would cost as follows,

$ $\frac{24.50}{1000}$ X 622 = $ 15.239

Rounded up to 2 decimal digits, we get $ 15.24

Hence, option 1 is the correct option.

SOLUTION 14 : $\frac{1}{3}$ X 2 $\frac{5}{8}$ and 0.24 / 0.2

$\frac{1}{3}$ X $\frac{21}{8}$ and $\frac{24}{100}$ / $\frac{2}{10}$

$\frac{7}{8}$ and $\frac{24}{100}$ X $\frac{10}{2}$ = $\frac{6}{5}$

Therefore, 7/8 and 1.2 are the right answers.

So, option D is the correct choice.

NOTE DOWN : Change the Division sign to Multiplication and do reciprocal at the same time.

SOLUTION 15 : Average show depth on Australia skifields =30 cm

By last week of the same month, the depth is reduced by 1/3 , so depth reduced = 30 X 1/3 = 10 cm

NOTE DOWN :

Here, Depth of snow = Height of snow

The new depth = 30 cm – 10 cm = 20 cm *Similarly, given that, the snow depth* for the whole month of August was 120 cm.

To Find : $\frac{\text{Snow depth in Australia (last week of August)}}{\textit{Snow depth in New Zealand (last week of August)}}$

= 20 cm / 120 cm = 1 / 6 or 1 : 6

Thus, option B is the correct choice.

SOLUTION 16 : Five-eighths of 24 plus 16 times 4 - 2 could be re-written as the following to make it equal to 47,

$\frac{5}{8}$ X 24 + 16 (4 – 2)

Now, lets ensure it is the correct one,

=> 15 + 16 (2) = 15 + 32 = 47 = 47 => Equation Satisfied , thus option B is the correct answer.

SOLUTION 17 : By triangle sum property, the addition of all the angles in a Δ is 180^0 .

Now, let us assume, the third angle in the triangle is x.

$\therefore \angle x + \angle 80 + \angle 70 = \angle 180$

So , by solving , $\angle x = \angle 30$

Also, by adjacent angle property,

$\angle a + \angle 2a + \angle x = \angle 180$

$\angle 3a = \angle 180 - \angle 30$

Thus, $\angle a = \angle 50$ is the correct answer.

SOLUTION 18 : By looking at the sequence carefully, the number pattern shows each number is divided by 3 every time we move forward,

Sequence is 27 9 3 1

To find : 7th term in the sequence

=> Fifth term in the sequence = 1 / 3

Similarly, sixth term = 1 / 3 ÷ 3 = $\frac{1}{9}$

Last, seventh term = $\frac{1}{9}$ ÷ 3 = $\frac{1}{27}$

Hence, Seventh term = $\frac{1}{27}$

SOLUTION 19 : To find out : Marian would need to cut down the cotton sheet from flag ; as the length and height of cotton flag would be 3 times her drawing size. We need to consider both length and breadth , both are 3 times each.

So, area she needs to cut down =(length X height)

=) (3 X 3) measurements = 9 times she will cut the cotton sheet.

Hence, option B is the correct option.

SOLUTION 20 : To calculate: Fuel consumption of large trucks in km/L

See the red point carefully in the graph, it appears to 60 L of fuel used and 150 kms of distance travelled.

Therefore, fuel consumption (in km / L) = 150 km/60 L

Thus, 2.5 km / L is a the correct answer.

SOLUTION 21 : Volume of water carrying in 1minute =6m X 3m X 3m => 54 m^3

∴ Volume of water in 1 second = 54 m^3 / 60 = 0.9 m^3

To calculate : Flow of water in L / sec

Let us begin by converting m^3 → into L

Note this that 1 m^3 = 1000 L

∴ 0.9 m^3 = 1000 X 0.9 L = 900 L

Hence, flow of water is $\frac{volume\ of\ water\ in\ litres}{time\ taken\ in\ seconds} = \frac{900\ L}{1\ s}$ or 900 L/s

Thus, option C is the correct option.

SOLUTION 22 : 13.2 X 8.7 = 114.84 ....1st equation

To calculate : 1320 X 87

Let us multiply equation 1 by 1000 , we get

13.2 X 8.7 X 1000 = 114.84 X 1000

$\frac{132}{10} \times \frac{87}{10} \times 1000 = \frac{11484}{100} \times 1000$

1320 X 87 = 114840

Thus, 114840 option C is the correct option.

SOLUTION 23 : Average height of 10 members = 1.68 m

Along with that, a new person joins with height = 1.79 m

To calculate : The new average height of the Farid's basketball squad.

∴ Average height = $\frac{total\ height\ of\ all\ 11\ members}{total\ number\ of\ members}$

= $\frac{1.68\ X\ 10+1.79}{11}$ = Option A is the correct option.

ADVANCED NUMBER

PERCENTAGES

1. (125 / 1000 X 100) % =(125 / 10) % = 12.5 %
2. To find alternative, solve 20 % which is 20 / 100 .
3. 25 / 100 X 100 = 25 % is the correct answer.
4. HINT : See Q 2, we just solved the percentage (50 %), 50 / 100 is the correct answer.
5. 62.5 % =(625 / 10) % = 625 / 10 X 1 / 100 = 625/1000
6. (75 / 100 X 100) % = 75 %
7. 80 % = 80 / 100
8. (90 / 100 X 100) % = 90 %

NOTE THAT :

We multiply by 100 when calculating the percentage and divide by 100 when need to find a fraction.

NUMBER PROBLEMS

SOLUTION : Let the number be x

Condition : If x is divided by 3 and add 4 to it, the result is 25.

∴ x/3 + 4 = 25 => x/3 = 25 – 4 = 21

∴ x/3 = 21 , thus x = 21 X 3 = 63 is the correct answer.

SOLUTION : Given that : 1/3 (AKASI 's class) = 9 students that were not born in Australia.

∴ Akasi's class = (9 X 3) = 27 total students

∴ The students that were born in Australia = Total number of students – Students that did not born in Australia

==> Students (born in Australia) = 27 – 9 = 18 .....1st equation

Similarly, other class in Akasi's school, the scenario remains the same. Means ,

total students = 27 ,

Born in Australia = 18 , ... 2nd equation

Not born in Australia = 9

∴ Using equation 1 & 2 , we obtain

Combined, total number of students in both classes that were born in Australia = 18 + 18 = 36

So, 18 is the correct answer.

SOLUTION : Total guests at Aldo's party = 120 (Given)

Let us assume that the party guests aged under 18 be 'a' .... 1st equation

∴ According to the given condition,

4 X guests (18 & over) = guests aged under 18

OR :=> 4 X guests (18 & over) = a

∴ guests (18 & over) = a / 4 ....2nd equation

∴ a + a / 4 = 120 => 5a = 120 X 4 = 480 [By taking L.C.M.]

∴ a = 480 / 5 = 96

∴ a = 96

Hence, party guests aged under 18 = 96 is the correct answer.

SOLUTION : Given that : 25 % of students in Arana's class have visited New Zealand.

∴ 75 % of students did not visit New Zealand = 24 (given)

Let us assume the total number of students be 'a'

∴ 75 / 100 X a = 24 => a = $\frac{24\,X 4}{3}$ = 32 students

Hence, the total number of students is equal to 32.

USING PERCENTAGES

SOLUTION 1 : Tip = \$ 15 (given that Bron gave it to waiter)

Also, it represents 8 % of the bill amount

To calculate : Bill amount

Let us assume Bill amount is 'a'

$\therefore$ 8 % of bill amount = 15 \$

$\therefore \frac{8}{100}$ X a = 15 \$

$\therefore$ a = $\frac{15 X 100}{8}$ \$ = $\frac{375}{2}$ \$ = \$ 187.50

$\therefore$ Bill amount = \$ 187. 50

SOLUTION 2 : Given that : Wei has a pile of 2950 programs approximately.

a. Number of boxes = Programs / Capacity = 2950 / 200

= 59 / 4 = 14.75 boxes

But the number of boxes cannot be in decimals , thus we use 15 boxes.

b. As it is 14.75, which is 14 + 0.75 = 14 + $\frac{75}{100}$ boxes used

$\therefore$ As it is clearly seen that 14 boxes are completely full, and 15th is 75 % full and 25 % empty.

$\therefore$ The answer is 25 % .

SOLUTION 3 : Given that : Special offer of 25 % makes buying :

3 T – shirts for \$ 45

$\therefore$ 1 T – shirt for \$ 15

To calculate : Original price per T – shirt

Let us assume the original price of 3 T-shirt be 'a'

$\therefore$ a X $\frac{25}{100}$ = 15 \$

$\therefore$ a = 60 \$

Thus, this is the price for all 3 T – shirts.

$\therefore$ ORIGINAL PRICE per T-shirt = 60 / 3 \$ = 20 dollars is the correct answer.

SOLUTION 4 : According to the Question,

\$ 2000 + 50 % of \$ 2000 = \$ 3000 which is indeed true.

Now, as we need to apply the same process to find 20 % of 3000 + 50 % of (20 % of 3000)

$$=> \frac{20}{100} X\ 3000 + \frac{50}{100} \left(\frac{20}{100}\ X\ 3000\right) => 600 + \frac{1}{2}\ X\ 600$$

==> 600 + 300 = 900

$\therefore$ 900 is clearly 30 % of 3000.

Hence, the answer is 30 % .

SOLUTION : Given the following conditions , Paul in ball-skills tests at the FUN PARK.

$\therefore$ Football winning chances = 25 % less success than playing basketball

Also, Cricket ball winning chances = 50 % (half) of success he got during playing football

Last but not the least, Snooker ball winning chances = 50 % MORE of success achieved during Cricket attempt

$\therefore$ Also, Basketball = 80 % success ...... (given in the table)

Now, Football winning chances = $\frac{25}{100}$ X 80 = 60

Similarly, Cricket ball winning chances = $\frac{50}{100}$ X 60 = 30

Similarly, Snooker ball winning chances = 30 + $\frac{50}{100}$

=> 30 + 15 = 45

So, the respective answers are 60 % , 30 % and 45 % .

FUN FACT :

Only Snooker ball has HIGHER chances, whereas on the other hand , football and cricket have LESSER chances.

THEREFORE, WE ADDED 50 % WHEN WE CALCULATED WINNING CHANCES FOR Snooker ball .

ADVANCED NUMBER

EXCHANGE RATES

SOLUTION : Maya brings = 30200 Yen ,165 Euros & 225 US dollars

She now wants to exchange these currencies into Australian dollars.

Given that : 1 AUD = 96.25 Yen = 0.79 EUR = 1.03 USD ......1st equation

Using equation 1, we obtain

HOT TIP :

Convert in AUD to the NEAREST CENT.

1 Yen = $\frac{1}{96.25}$ AUD , 1 EUR = $\frac{1}{0.79}$ AUD , 1 USD = $\frac{1}{1.03}$ AUD ;

1. 30200 Yen = ($\frac{30200}{96.25}$) AUD = 313.766 AUD = 313.77 AUD
2. 165 EUR = (165 X $\frac{1}{0.79}$) AUD = 208.86 AUD
3. 225 USD = ($\frac{1}{1.03}$ X 225) AUD = 218.446 AUD = 218.45 AUD

NUMBER STEPS

HOT TIP :

Ascending order is an Increasing order , that is , from SMALLEST to LARGEST !!

1. $\frac{2}{3}$ X 0.33 = $\frac{2}{3}$ X $\frac{33}{100}$ = $\frac{11}{50}$ = 0.22
2. 18 % / 3 = $\frac{18}{100}$ X $\frac{1}{3}$ = 0.06
3. 20 % of 2.5 = $\frac{20}{100}$ X 2.5 = $\frac{1}{5}$ X $\frac{25}{10}$ = $\frac{1}{2}$ = 0.50
4. $\frac{1}{2}$ of 70 % = $\frac{1}{2}$ X $\frac{70}{100}$ = $\frac{7}{20}$ = 0.35

Now, let us arrange them in ASCENDING ORDER , we get

1. 18 % / 3
2. $\frac{2}{3}$ X 0.33
3. $\frac{1}{2}$ of 70 %
4. 20 % of 2.5
5. 24 X 125 % = 24 X $\frac{125}{100}$ = 24 x $\frac{5}{4}$ =30
6. 75 % of 16 = $\frac{75}{100}$ X 16 = $\frac{3}{4}$ X 16 = 3 X 4 = 12
7. 27 / 0.75 = 27 / $\frac{75}{100}$ = 27 x $\frac{100}{75}$ = 27 X $\frac{4}{3}$ = 36
8. $6\frac{1}{2}$ / 0.25 = $\frac{13}{2}$ / $\frac{25}{100}$ = $\frac{13}{2}$ X $\frac{100}{25}$ = $\frac{13}{2}$ X 4 = 26

Now, let us arrange them in DESCENDING ORDER , we get

5. 27 / 0.75
6. 24 X 125 %
7. 6 $\frac{1}{2}$ / 0.25
8. 75 % of 16

HOT TIP :

Descending order is a Decreasing order , that is , from LARGEST to SMALLEST !!

NUMBER SENTENCES

SOLUTION 1 : Number sentence for the number of extra golf balls and solution is as follows :

=> 7 – (2 X 2) – 1 + 2 = 4 is the correct answer.

SOLUTION 2 : According to the question ,

Boats that are large = 2

Boats that are small = 1

2 LARGE BOATS : 1 SMALL BOAT

Carry ↓ Carries ↓

(18 people each + 2 crew in each boat)

(14 people + 1 crew)

Total 5 crew members(4 from large boat and 1 from small boat) are not DIVERS, so they won't be counted, hence we subtract them.

$\therefore$ 18 X 2 – 2 X 2 + 14 -1 = 45 OR 2 (18 – 2) + 14 -1 = 45

Is the number sentence to represent the given condition and total number of divers are 45 that were taken out to the reef.

ADVANCED MEASUREMENT AND SPACE

VOLUME AND CAPACITY

SOLUTION 1 : Given that : Dee's backyard spa pool is rectangular in =>

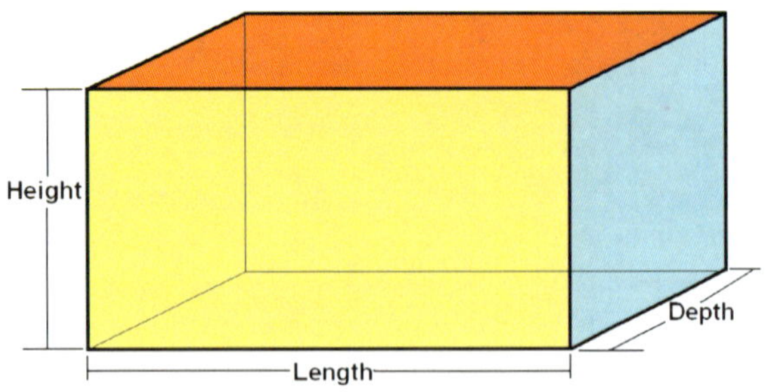

RMEMBER THAT : If a pool is rectangular in shape means each face is rectangular; it will make a CUBOID. It is so because LENGTH , BREADTH AND HEIGHT are different to each other. If all 3 dimensions were same , it would have been a CUBE.

Here, Length l = 2.4 m ; Breadth b = 1.8 m & Height h= 1.2 m ... (GIVEN)

Internal volume of empty pool = l X b X h

=> 2.4 m X 1.8 m X 1.2 m => $\frac{24 \ X \ 18 \ X \ 12}{1000}$ = 5.184 m^3

==> 5.184 m^3 would be 5.2 m^3 ....(to the nearest one decimal place) is the correct answer.

SOLUTION 2. To find : capacity of pool in kl to 2 decimal places.

Also, the pool is filled up with depth (height) = 1080 mm

= 1.08 m [∵ 1 m = 1000 mm]

∴ capacity of filling the water in the pool = l X b X h

= (2.4 X 1.8 X 1.08) m^3 => 4.665 m^3

= 4.67 m^3 up to 2 decimal places.

FUN FACT :
1 kL
= 1 cubic metre
= 1 m^3 !!

As we need to find capacity in kilo litre (kL) , so we need to convert it into kilo litre .

∴ 4.67 m^3 = 4. 67 kL is the correct answer.

SOLUTION 3. Flow rate of the hose = 12 L / min = (12 X 60) L / hour

To calculate : Time to fill up the pool that has a depth of 1080 mm or 1.08 m

Hence, Time = $\frac{Capacity}{Flow \ rate} = \frac{4.67 \ kL}{720 \ l \ (in \ hours)} = \frac{4.67 \ X \ 1000 \ L}{720 \ l \ (in \ hours)}$

=> 6.48 hours = Time : converting it to ONE decimal place,

Time = 6.5 hours is the correct answer.

∵ 1 *kL* = 1000 *l*

CIRCLE PROPERTIES

SOLUTION 1 : Chord = b (line segment joining two points on a curve)

SOLUTION 2 : Radius = c (the distance from center to the boundary)

SOLUTION 3 : Circumference = a (total length of boundary of circle)

Note : Have a glance at the diagram carefully .

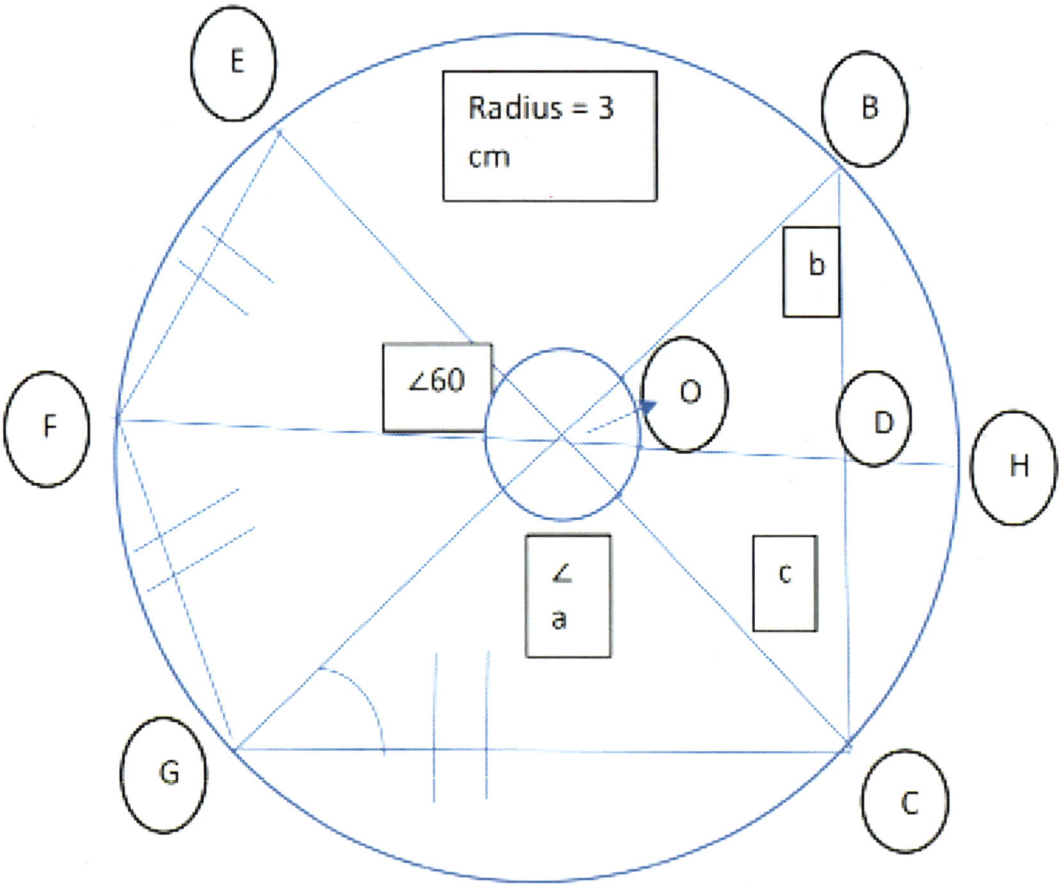

We will name certain points in the diagram , see them

Radius = 3 cm & O → *centre of the circle*

As ∠EOF = ∠COD (∵ Vertically Opposite Angles)

Also, OF = OG = RADIUS = 3 cm

If sides of the triangle are equal ,

$\therefore$ 2 angles are also equal. ($\because$ Isosceles Δ theory)

Thus, $\angle OFG = \angle OGF = \angle 60$

$\therefore \angle 60 + \angle FOG + \angle 60 = \angle 180$ ($\because$ Δ sum property)

$\therefore \angle FOG = \angle 60$

Now, $\angle FOG + \angle a + \angle EOF = \angle 180$ (*Adjacent angle*)

=> $\angle 60 + \angle a + \angle 60 = \angle 180$ ($\because \angle EOF = \angle 60$ => given)

$\therefore \angle a = \angle 60$ is the correct answer.

SOLUTION 4 : Also, $\angle a = \angle EAB = \angle 60$ => Vertically opposite angles

SOLUTION 5 : Similarly, $\angle BOD = \angle 60$ => Vertically opposite angles

As we could see clearly, Δ BOC is an isosceles triangle because OB = OC = radius of the circle

$\therefore \angle b = \angle c$ ....1st equation

Also, for Δ OBC,

$\angle o + \angle b + \angle c = \angle 180$ ($\because$ Δ sum property)

$\therefore \angle 120 + \angle b + \angle b = \angle 180$ (*here angle o is 120 degrees* ($\angle 60 + \angle 60$) *for this triangle* & $\angle b = \angle c$ using equation 1)

$\therefore 2 \angle b = \angle 60$

Thus , $\angle b = \angle 30 = \angle c$

SOLUTION 6 : $\angle c = \angle 30$ (*as proven above*)

SOLUTION 7 : The short chord would be EF = FG = GC

These all 3 chords are equal as given in the question.

As we know that in Δ EOF,

$\angle$ *OFE* = $\angle$ *OEF* (*Isosceles triengle theory*)

Also, $\angle$ *OEF* + $\angle$ *OFE* + $\angle$ *EOF* = $\angle$ 180 (*Triangle sum property* && $\because$ $\angle$ EOF = $\angle$ 60 => given)

=> 2 $\angle$ *OFE* = $\angle$ 180 − $\angle$ 60 = $\angle$ 120

$\therefore$ $\angle$ *OFE* = $\angle$ 60 = $\angle$ *OEF* = $\angle$ EOF

If all angles of a triangle are equal, it means it is an Equilateral triangle. So, all the sides of an equilateral Δ are equal.

Thus, FE = OF = OE = RADIUS = 3 cm

Therefore, EF = FG = GC = 3 cm is the length of all chords.

SOLUTION 8 : Circumference of circle = 2 π r

C = 2 X 3 X r = 6 X r is the correct answer when a whole number multiplies the radius of the circle.

FUN FACT :

π = 3.14 *or 3 as a Whole* NUMBER !!

SOLUTION 9 : As Circumference of circle = 2 π r

C= Diameter X π

So here, the circumference is equal to *π multiplied* by the diameter of the circle. (diameter = 2 X radius)

$\therefore$ The number *π* is equal to 3.14 , but we need a constant whole number (means no decimals included) that gives us 3.

Hence, 3 is the correct answer.

ADVANCED MEASUREMENT AND SPACE

3-D SPACE

SOLUTION : From the given figures, the correct top view of the given 3-D shape will be option b . Because that is EXACTLY what we see when we look at something directly above 3-D shape.

So, option b is the correct option.

SOLUTION 1 : Clearly , it is a 3-dimensional solid object in which the 2 identical faces are in the shape of a triangle , also connected by 3-rectangular faces.

Therefore, it is a TRIANGULAR PRISM .

SOLUTION 2 : As seen clearly from the figure , a triangular prism has 9 edges in total.

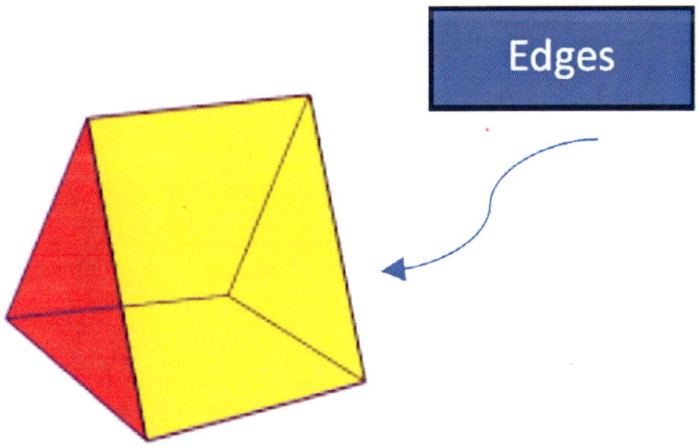

2–D TRANSFORMATIONS

SOLUTION 1 : N , S , Z are the letters that could be rotated 180 degrees and would never change (means do them upside down and you will be able to read them as they are) . ....[Refer to below table]

SOLUTION 2 :

A	B	C	D	E
F	G	H	I	J
K	L	M	N	O
P	Q	R	S	T
U	V	W	X	Y
Z				

Vertical

Horizontal

Look at the above table carefully,

Rotate the alphabets horizontally, and the below one's wont's change :

A M T U V W Y

TIMETABLES AND TIME ZONES

Given that HANA is planning a trip from Sydney to Perth.

Train time to Sydney airport by fast train = 1 h 37 min

Train time to Sydney by slow train = 2 h 23 min

SOLUTION 1 : If Hana can arrive no later than 10:15, she must catch 07:44 slow train because after 2 h 23 min , the time would be 10:07 which is acceptable as it is before 10:15 . So, she won't be late.

Therefore , Option 1 is the correct option.

> POINT TO REMEMBER : Putting other 2 options would make her arrive late by a couple of minutes .

SOLUTION 2 : As discussed in the very last question , the most suitable train would reach at 10:07.

SOLUTION 3 : To calculate : Sydney to Perth flying time.

=> As the plane departs at 12:05 Sydney daylight time , means it is 3 hours ahead of Perth time.

Therefore, Perth time when the plane departs in Sydney would have been 09:05 . As the plane finally lands in Perth at 14:00 local time, ∴ difference between (09:05 & 14:00) must be the total time taken by the plane to reach Perth, which is 04:55 04 4 hours 55 minutes.

Hence , 04 hrs. 55 minutes is the correct answer.

SOLUTION 4 : Perth time (when the plane departs) = 16:20

∴ It takes 4:05 hours to Sydney ......(Given)

Must have been (16:20 + 4:05 hours) Perth local time => 20:25 hours Perth time

∴ Converting this into Sydney time , add 2 more hours as Sydney is 2 hours ahead than Perth.

=> (20:25 + 2 hours) Sydney local time

∴ 22:25 is the time the plane should take to land in Sydney and hence is the correct answer.

SCALE AND POSITION

SOLUTION : Scale = $\frac{1}{200} = \frac{Drawing\ of\ Emilia's\ house}{Garden}$ ....(Given)

∴ Drawing measures 20 cm as the length of block of land. .... (given)

To calculate : Distance that 20 cm on the drawing represent on the ground .

∴ 20 cm X 200 times = 4000 cm = 40 m [∵ 1 m = 100 cm]

Hence, 40 metres is the correct answer.

ADVANCED ALGEBRA , PATTERNS AND DATA PATTERNS

SOLUTION : Given : First hexagon is made up of 6 rods

2nd hexagon is made up of 11 rods

Similarly, 3rd hexagon is made up of 16 rods

Conclusion : Every new hexagon means adding every 5 rods.

To find : Shape number using 61 rods

As difference between 61 & 16 = 45

∴ 45 shape number / 5 rods = 9th shape number after 16 rods

∴ 9 + 3 = 12

[We added 3 cause there were already 3 shapes till 16 rods]

∴The answer is 12 as 12 shape number could be made using 61 rods.

SOLUTION : After analyzing the table very cautiously, we get

1. For 1 ; there is a reducing difference of 2 , 4 , 6 , 8 ; so, the 6th number would be 102 – 10 = 92

2. For 2 ; multiply the previous number and subtract 2 ; we will get (163 X 3) – 2 = 489 – 2= 487

3. For 3 ; the number pattern could be adding previous number to 3 times of it, for instance

   =>   2 (3) + 2 = 8

   Similarly, 8 (3) + 2 = 26

   And similarly , for the sixth number,

   242 (3) + 2 = 726 + 2 = 728

4. For 4 ; the number is SIMPLY multiplying by $\frac{1}{10}$, we get 500 + $\frac{1}{10}$ × 500 = 500 + 50 = 550 which is the 2nd number ,

   Similarly, 550 + $\frac{1}{10}$ × 550 + 55 = 605

Similarly, moving to the 6th number could be as follows.

$732.05 + \frac{1}{10} \times 732.05 = 732.050 + 73.205$

=> 805.255 which is the correct answer.

SOLUTION : Given that : Dimensions of first cube = 1 cm X 1 cm X 1 cm

Also, second cube 's , third cube's , fourth cube 's & fifth cube's dimensions have been INCREASED by 1 cm of length , breadth & height (depth) .

Thus , Volume series for all individual cubes ; we get

1 cm X 1 cm X 1 cm = 1 cm^3

2 cm X 2 cm X 2 cm = 8 cm^3

3 cm X 3 cm X 3 cm = 27 cm^3

Similarly, 4 cm X 4 cm X 4 cm = 64 cm^3

Now , finally the fifth number in this pattern ,

5cm X 5 cm X 5 cm = 125 cm^3

==> Volume of the fifth cube in cm^3 is 125 which is the correct answer.

SAMPLING AND PREDICTIONS

Given that : Diego purchased large bag of jellybeans where jellybean company mixes 5 colors of beans in equal numbers.

SOLUTION 1 : As Diego takes out 50 beans randomly, therefore the number of black beans could be one color out of 5 , so a total of 10 beans may be black according to the given statement.

SOLUTION 2 : No , it seems like out of 150 beans , Diego only gets 16, which is very less than normal. So, it is not an accurate sample of the company's product.

SOLUTION 3 : See the Q2. , out of first 150 beans , Diego counts 16 black beans .

Therefore, out of next 100 beans left , there must be 50 – 16 = 34 black beans because (150 + 100) 250 beans must have exactly one-fifth of the black beans as well.

So , the answer is 34.

SOLUTION 4 : The prediction is not related to probability in this case , so even if his prediction about the third batch are wrong , it will not mean Diego's probability calculations were wrong.

Thus, the answer is No.

SOLUTION 5 : Yes, because black beans must be exactly $1/5^{th}$ of the total beans .

=> Total beans = 50000

$\therefore$ Black beans = $\frac{1}{5}$ X 50000 = 10000

$\therefore$ Black beans according to question is 10000 which is a correct answer.

USING INVERSE OPERATIONS

GIVEN : Letters A and H are numbers.

1. 25 + A = 97

   A = 97 – 25 = 72 is the correct answer.

2. (B + 9) / 4 = 4

   (B + 9) = 4 X 4 = 16 [By cross multiplication]

   B = 16 – 7 = 9 is the correct answer.

3. $2^3 + C = 27$

   $C = 27 - 2^3 = 27 - 8 = 19$

Thus, the value of C is 19 which is the correct answer.

4. 30 – D = 56 / 7 =8

   30 – D = 8

   30 – 8 = - (- D)

   22 = + D

   ∴ The value of D is 22 .

5. E X (13 – 8) = 45

   E X 5 = 45

   E = 45 / 5 = 9

   So, E = 9 is the correct answer.

FUN FACT :

If any number changes its side, moving from LHS (LEFT HAND SIDE) TO RHS (RIGTH HAND SIDE) OR VICE-VERSA ; then,

1. + converted into –
2. – converted into +
3. ÷ converted into X
4. X converted into ÷

6. 2 X F + 5 X 6 + 2 = 8^2

   2F + 30 = 64 – 2

   2F = 62 – 30 = 32

   F = $\frac{32}{2}$ = 16 is the correct answer.

7. $\frac{G}{3} + \frac{G}{2} = 10 \Rightarrow \frac{2G+3G}{6} = 10$

   5G = 10 X 6 = 60

   G = 12 is the correct answer.

8. 7 – 10 = (H + 5)

   - 3 – 5 = H

   Hence, H = - 8 is the correct answer.

ADVANCED ALGEBRA , PATTERENS AND DATA

DATA – CRITICAL EVALUATION

Note : Study the line curve graph carefully. It is (HOURS SPENT STUDYING PER WEEK) TO (AGE OF STUDENTS) (IN YEARS) TYPE OF GRAPH.

SOLUTION 1 : No, because 0 is not used for both the data used in the graph. (X – axis or Y – axis)

SOLUTION 2 : No, because 0 is not used for both the data used in the graph. (X – axis or Y – axis)

SOLUTION 3 : No, the fact that the students are overworked is not mentioned in the graph or in the question itself.

SOLUTION 4 : As we cannot conclude the data till age 17 because the given graph does not DISCLOSE / DISPLAY this specific information , so we con not really predict about this question.

Therefore , the answer is no.

SOLUTION 5 : Yes, as the line curve graph shows the data / information upwards towards y- axis ; so, this could be easily overcome by using column graph that could plot age groups up to 17.

SOLUTION 6 : To plot 3 age groups, we require EXACTLY 3 distinct columns . So, the answer is 3.

SAMPLE TEST

SOLUTION 1 : Given that : A cube has 27 small 1 cm^3 blocks.

Also , Linda adds 9 more blocks to make a new front , and 12 blocks to make a new side.

To calculate : The number of blocks she will add to make a new TOP.

As we know that , by adding 9 blocks in the front would be 4 X 3 X 3 (length X breadth X height)

Moreover, after adding 12 more blocks at side, the dimensions would be 4 X 4 X 3 .

NOTE DOWN :

Here just breadth is increased as we added blocks on the side , no height is increased as we did not yet add any blocks on the TOP.

Now, (4 X 4) blocks (length X breadth) will be required to make a new TOP as height is entirely dependent on length and breadth in this case.

∴ 4 X 4 = 16 blocks is the correct answer.

Thus, Option 1 is correct.

SOLUTION 2 : 1st number = 3

2nd number = 6

3rd number = 5

4th number = 10

Rule : Multiply a previous ODD number by 2 and subtract 1 from a previous EVEN number.

To calculate : 8th number in the series .

Fifth number = Even number – 1 = 10 – 1 = 9

Sixth number = Multiply odd number by 2 = 9 X 2 = 18

Similarly, 7th number = 18 – 1 = 17

Finally, 8th number = 17 X 2 = 34

∴ Option 4 is the correct option.

SOLUTION 3 : Thomas would catch the train at Cosgrove station at 8:38 am to reach Barnley station before or at 9:10 am ; which in this case is 9:03 am. If he catches other 2 trains ; he will be either late or way too early to get to the Barnley station.

Thus, 8:38 am is the correct answer.

SOLUTION 4 : I) $22\,\% = \frac{22}{100} = \frac{11}{50} = 0.22$

II) $\frac{1 X 0.3}{4 X 0.3} = \frac{1}{4} = 0.25$

III) $\frac{5}{30} = \frac{1}{6} = 0.16$

IV) 0.2

Hence, out of all 4 options , the smallest number is 0.16 .

Thus, Option 3 is the correct option.

SOLUTION 5 : Isosceles Δ is a triangle whose exactly 2 sides are equal, and 2 angles are equal.

To calculate : Angle a + Angle b

Sol : See the figure carefully , in the largest Δ, the sum of all angles is 180^0.

[∴ By Δ sum property]

=> $\angle a + \angle b + (\angle a + \angle b) = \angle 180$

=> $2 \angle a + 2 \angle b = \angle 180$

=> $2 + (\angle a + \angle b) = \angle 180$

==> $(\angle a + \angle b) = 180 / 2 = \angle 90$

Thus, $(\angle a + \angle b) = \angle 90$ which is the correct answer.

SOLUTION 6 : 2 % of 15750 needs to have the sample numbering according to the question .

Thus, 15750 X $\frac{2}{100}$ = 315 people

Now, as per the question , if the majority of sample (out of 315 people) agrees , the COUNCIL would be building a walking and cycle track inside the reserve.

$\therefore$ $\frac{315}{2}$ people = 157.5 ≅ 158 *people* as number of people cannot be in decimals . So , it must be a WHOLE number.

Hence, 158 people need to agree for the MAJORITY WIN.

HOT TIP : A majority is referred to anything that is in EXCESS of 50 % !!

Here, ≅ sign is a sign of APPROXIMATELY.

SOLUTION 7 : Distance from Graydon to Mosely = 11.3 km ≅

To calculate : Approximate distance between Mosely to Elmore .

As is clearly observed from the given figure, the path from Graydon to Mosely and from Mosely to Elmore is diagonally crossing the GRID.

It pretty much seems that distance from M to E is $\frac{3}{4}^{th}$ approximately of distance from G to M.

Thus, distance (M to E) = $\frac{3}{4}$ X 11.3 km = 8.47 km

And it is closest to Option C = 8.5 km is the correct answer.

SOLUTION 8 : As Jae buys skateboard for dollars = 68 $ at a discounted price of 15 % ,

$\therefore$ The price he paid ($ 68) is 85 % of 80 = $\frac{85}{100}$ X 80

=> 17 X 4 = $ 68 which matches the discounted price.

Hence, option B is the correct answer.

SOLUTION 9 : Total number of students in Nadia's class = 26 students

Students that had black hairs and dark eyes = 15 students

Also, the students that had blond hair and green/ blue eyes = 11 students

To find : Probability a student has either black or blond hair will be as follows :

$$\Rightarrow \frac{\textit{Students with black hairs} + \textit{Students with blond hairs}}{\textit{Total number of students}} = \frac{15+11}{26}$$

=> P (chance) = $\frac{26}{26} = 1$

Thus , Option B is the correct option.

NOTE DOWN : The question asks about hairs NOT about eyes ; and every student has either black or blond hair , as there is no other hair color.

SOLUTION 10 : K = - 5 (given)

Equation => $3^3 - 32 = K$

$\therefore 27 - 32 = K$

$-5 = K$

K = - 5 ....1st equation

Also, P = K + 15 ....2nd equation

Therefore, putting equation 1 in equation 2, we get

P = -5 + 15 = 10

So, P = 10

Hence, Option C is the correct option.

SOLUTION 11 : As $\frac{5}{11} = 0.45 = \frac{45}{100} = 45\%$

Thus , Option D is the correct option.

SOLUTION 12 : As observed clearly in the BAR graph (FEB – APR) would be Autumn / Summers and (JUN - AUG) be Winters. Both are represented by distinct colors that is bright blue color and dark blue respectively.

Now, Number of students that chose swimming are 120. Similarly, the number of students that chose football are 60.

∴ Their difference = 120 – 60 = 60 students are those that chose swimming in Autumn / Summers than football in the winter.

Hence, 60 is the correct answer.

SOLUTION 13 : The given line is cutting the circumference of circle in 2 different arcs.

Therefore, it is a SECANT.

SOLUTION 14 : As seen clearly in the picture , 8 diamonds are formed in a pattern by combining 24 bars.

Also, the table displays the number of bars along with number of diamonds formed using those bars.

To calculate : Number of bars required to form 20 diamonds

Using the right-hand side FUN FACT ;

Number of diamonds (11) = (24 + 8) bars

=> 32 bars

Similarly, number of diamonds (14) = (32 + 8) bars

=> 40 bars

Similarly, number of diamonds (17) = (40 + 8) bars

=> 48 bars

Finally, to obtain 20 diamonds = (48 + 8) bars

=> 56 bars

Hence, 56 bars would be used IN ORDER TO get 20 diamonds.

FUN FACT :

A difference of 3 diamonds is OBSERVED by adding 8 extra bars.

SOLUTION 15 : In the given equation as follows ;

$(8^2 - 5 \text{ X } 6) - 2 / 2 = N^2$

To find : Value of N

=> $(64 - 30 - 2) / 2 = N^2$

=> $32 / 2 = N^2$

=> $16 = N^2$

Thus, N = 4 is the correct answer.

SOLUTION 16 : Given: 1 AUD = 1.3 SGD

REMEMBER THAT :

AUD = Australian dollar

SGD = Singapore dollars

$\therefore \frac{1}{1.3}$ AUD = 1 SGD .... 1st equation

& 125 AUD = 125 X 1.3 SGD = 162.5 SGD

Thus, Kwan got 162.5 SGD.

Now as he spent 120 SGD , he is left with

162.5 SGD – 120 SGD = 42.5 SGD

Hence, converting the remainder amount back into AUD, 42.5 X $\frac{1}{1.3}$ AUD = 32.69 AUD

Thus, Option D is the correct option.

SOLUTION 17 : $\frac{10 \, X \, A}{2} \, X \, \frac{A}{15} = 12$

To calculate : Value of A

=> $5A \text{ X } \frac{A}{15} = 12 = 5(A^2) = 12 \text{ X } 15$

$= 5(A^2) = 180 \Rightarrow (A^2) = \frac{180}{5} = 36$

Therefore, A = 6 is the correct answer.

SOLUTION 18 : Correct right-side view means if we stand at right to given house, we will be able to see the HUT shape & nothing else including CHIMNEY. Figures B , C & D have got CHIMNEYS . ONLY option A does not display because going at R.H.S. (RIGHT HAND SIDE) of the house, we MUST not see the small door house connected.

Thus, Option 1 is a correct option.

SOLUTION 19 : Given that :

Mumbai time = 4 hours 30 minutes behind of Melbourne's time

$\therefore$ Mumbai time when it is 18:00 in Melbourne would be as follows :

=> 18:00 – 4 hours 30 minutes => 6pm – 4 hours 30 minutes => 1:30 pm

[$\because$ 1800 = 6*pm*]

Thus, Option 1 is the correct option.

SOLUTION 20 : Given :

Kym and Shelley took photos on shared camera = 65 ....1st equation

Assume : Let the number of photos taken by Kym be a

$\therefore$ Photos taken by Shelley = a + 17

Also, using equation 1 , we get

$a + (a + 17) = 65$

$\therefore\ 2a = 65 - 17 = 48$

$\therefore\ a = 24$

Hence, the number of photos taken by Kym (a) = 24 photos

Moreover, number of photos taken by Shelley a + 17 will be equal to 24 + 17 = 41 photos

SOLUTION 21 : After rotating the given figure to 180^0, we get

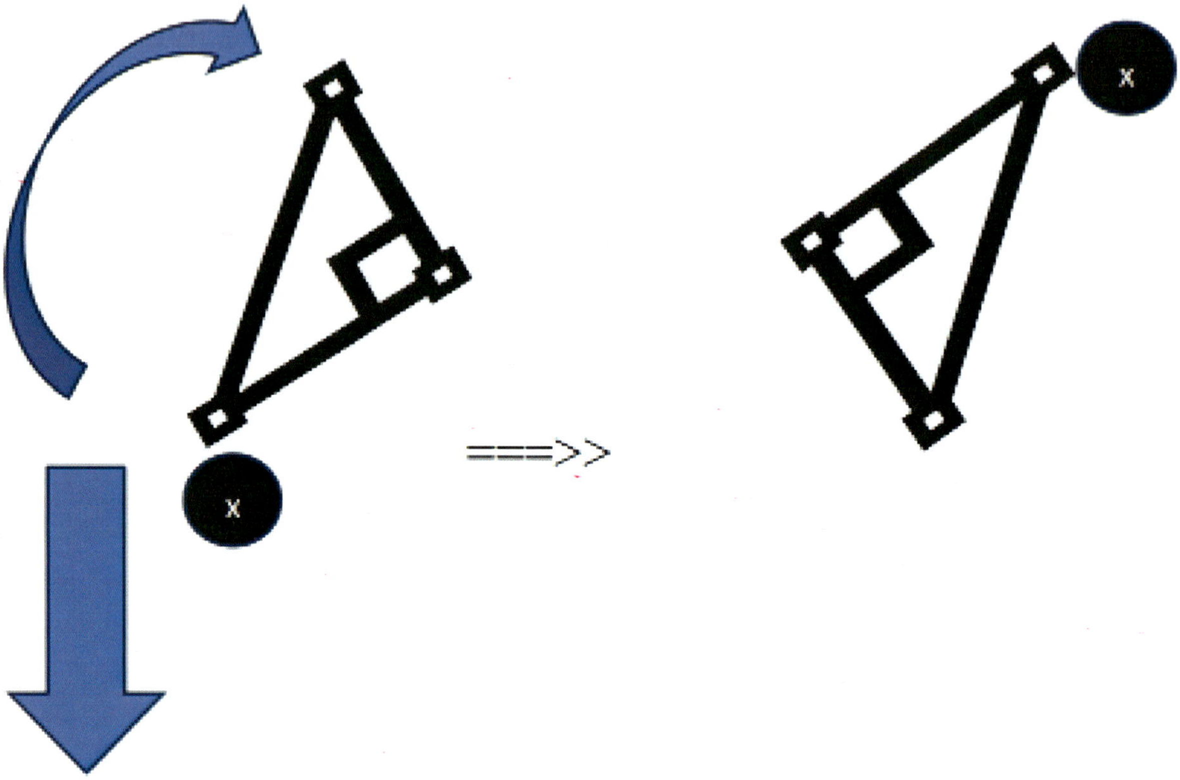

Rotate it 180^0 around x ; and we obtain the Final figure .

Thus, Option C is a correct option.

SOLUTION 22 : Definition : A rectangular prism is a 3 – D solid figure . It has , in total, 6 faces (rectangular).

Also, the TOP & BASE of the rectangular prism are always a rectangle. It also looks exactly like a CUBOID when folded up.

Since the definition says , the given figure could have FACE 3 as a base & Face 5 as on its TOP 2 & 5. More or less, FACE 2 & FACE 4 will be sides opposite to each other.

Similarly, FACE 1 & FACE 5 are also sides of the PRISM and would be opposite to each other.

Thus, the answer is FACE 6.

Made in United States
North Haven, CT
05 November 2026

21766151R00051

NAPLAN

HOW TO CLEAR YOUR **NAPLAN WITH EASE**

CLEAR SOLUTIONS

Get ready to conquer Naplan Year 7 Math with this comprehensive guide by Mayank Kumar. It doesn't just provide answers; it unravels the exact steps and reasoning behind each solution.

STEP-BY-STEP GUIDANCE

Navigate through the toughest Naplan challenges effortlessly. Each solution is presented with a step-by-step approach, ensuring you understand the "how" and "why."

PROBLEM-SOLVING MADE EASY

Say goodbye to confusion and hello to clarity. Mayank Kumar's book simplifies complex math problems, making them a breeze to tackle.

THE ULTIMATE MATH COMPANION

Whether you're a student, parent, or teacher, this book is your ideal sidekick for mastering Year 7 Naplan Math.

INTRODUCING

MAYANK KUMAR'S

YEAR 7 NAPLAN MATHS SOLUTION BOOK

Simplify your math journey with **Mayank Kumar's Year 7 Naplan Maths Solution Book**. Don't just solve problems; **understand them!**

Author Coach: Shoma Mittra
Cover Design: Manish Saha

ISBN 9780646889894
90000
9 780646 889894